THE ATMOSPHERE OF HEAVEN

THE
ATMOSPHERE
OF HEAVEN

THE UNNATURAL EXPERIMENTS OF
DR BEDDOES AND HIS SONS OF GENIUS

Mike Jay

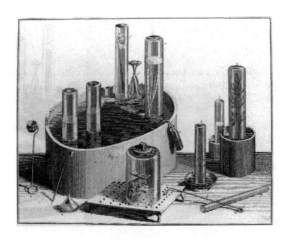

YALE UNIVERSITY PRESS
NEW HAVEN AND LONDON

'You are a philosopher', said the lady, 'and a lover of liberty. You are the author of a treatise called "Philosophical Gas; or, a Project for the General Illumination of the Human Mind".'

Thomas Love Peacock, *Nightmare Abbey* (1818)

As the births of living creatures are at first ill-shapen, so are all Innovations, which are the births of time.

Francis Bacon, *Of Innovations* (1612)

For information about this and other Yale University Press publications, please contact:
U.S. Office: sales.press@yale.edu   www.yalebooks.com
Europe Office: sales@yaleup.co.uk   www.yaleup.co.uk

Set in Arno Pro by IDSUK (DataConnection) Ltd.
Printed in Great Britain by TJ International Ltd, Padstow, Cornwall

Library of Congress control number: 2009924557

ISBN 978-0-300-12439-2

A catalogue record for this book is available from the British Library.
10 9 8 7 6 5 4 3 2 1

Contents

Illustrations

Bastille Day 1791

On the evening of Thursday 14 July 1791 Joseph Priestley was at home in Fair Hill, his three-storey town house on the outskirts of Birmingham, and was sitting down after supper to a game of backgammon when he heard a violent knocking at the door. Some young men, breathless from running, announced that a 'great mob' was on a destructive rampage through the town.[1] This had begun by venting its fury on the venerable and visible institutions of the dissenting community, the New and Old Meeting Houses, which it had been pelting with stones and was now tearing down and putting to the torch. To judge by the clamour of voices, its next target was Priestley: his house, his laboratory and the great chemical philosopher himself.

Though characterised by his enemies as a fiery iconoclast full of venom against the authority of the established church and monarchy, and routinely caricatured as Satan himself, Priestley in person presented a mild and diffident figure, famously likened to a woodpecker for his beaky nose, hopping gait and constant stammering chatter. His response to the emergency, however, was calm and deliberate. He hid a few papers and valuables, then jumped into a chaise and removed himself to a friend's house a mile out of town. From here, on rising ground, he could see the city below him washed in bright moonlight, the two meeting houses haloed in flame and evidently burning to the ground; and he could hear 'every shout of the mob' as it began to converge on his house with staves and wrecking bars. Once the rioters had surrounded it and smashed their way inside, they were, according to reports, 'incensed' to find Priestley gone, as they 'certainly meant to sacrifice him' (by some accounts, they had intended to roast him alive, and had even brought along a spit for that purpose). They compensated by cobbling together an effigy of him, which they hanged 'in the most ignominious manner' and torched 'amid the shouts and acclamations of near ten thousand people'.[2]

Priestley, who had assumed that the potential danger was to his windows and perhaps in the worst case to his person, watched and listened in horror as hundreds of his fellow-citizens invaded his house and set about destroying what he would later call 'the most truly valuable and useful apparatus of philosophical instruments that perhaps any individual, in this or any other country, was ever possessed of'.[3] The collection of equipment that Priestley, Britain's foremost experimental chemist, had accumulated over the years was as idiosyncratic and potent as the work to which he had put it. The world's finest laboratory glassware, blown and cut to order and provided free by specialist manufacturers in London, cohabited with conveniently sized beer glasses from Mrs Priestley's kitchen; the earthenware baths and troughs, and the china crucibles and retorts, were custom-made from Cornish clay by his friend Josiah Wedgwood. The legendary glass lens, a foot in diameter, with which he had heated mercuric oxide to isolate oxygen in 1774, was never seen again after the riots. Yet despite the inferno they had made of the meeting houses, the crowd proved unable to set fire to Fair Hill: instead, they virtually dismantled the entire house by hand, one of their number being killed by falling timber in the process. Priestley could hear raised voices offering two guineas for a lighted candle, and the next day he learned that 'much pains was taken, but without effect, to get fire from my large electrical machine, which stood in the library'.[4]

By 3 a.m. the house had been razed. Priestley, deep in shock and aware that nothing could be done until morning, was just settling down to attempt some sleep when news arrived that the mob had discovered his whereabouts and was advancing towards him. He remounted the chaise and, as dawn broke, took sanctuary in the house of another friend in the outlying town of Dudley. The following day, he learned that the rioters had picked up his scent once more; he waited for nightfall before slipping away on horseback and riding hard across country in the direction of the Welsh border to Bridgnorth.

Meanwhile, the rioting in Birmingham had spread. The city centre was a chaos of fire, broken glass, looting and mass drunkenness, walls and buildings being daubed with the slogans 'Church and King forever' and 'Damn Priestley'. But amid the chaos there were signs of concerted organisation. The objects of the mob's fury had been carefully selected to target the dissenting community, and gangs of muscular wreckers from the surrounding Black Country had appeared on cue to ratchet up the damage. It was also conspicuous that the local magistrates were taking no action to protect the dissenters' property: several of them were holed up in a tavern in the town centre, steadfastly ignoring pleas to intervene. It was only by the weekend, when Peck Lane prison had been broken into and the prisoners liberated, and the mob was beginning to eye up the estates of the local landowning gentry, that the Riot Act was read and the county militia summoned. Over the next few days four thousand dragoons streamed into the centre of town, requisitioning inns and billeting themselves among the populace until public order was restored.

Once the riot had taken hold, it was, like most riots in eighteenth-century Britain, unclear precisely what the mob was protesting against. Many in the crowd were heard shouting 'No Popery', a platform on which Priestley himself was a conspicuous campaigner. But its underlying causes were clear enough. The spark had been a Bastille Day dinner given by the Revolution Society, a civic organisation which for a century had commemorated the liberty won by the British people in the Glorious Revolution of 1688 and which was now also celebrating that more recently won by the French people from their despotic monarchy. Priestley was a member of the Revolution Society, though not present at the dinner on this occasion. It had been been advertised in advance with inflammatory and seditious handbills which the society claimed were malicious forgeries; partly in response to this provocation the dinner had been a self-consciously moderate affair, with toasts conspicuously drunk to the king and constitution. Rumours of an impending riot had been circulating for days, against a background of tensions that had been simmering for years between Church and King loyalists and a dissenting community increasingly vocal in its public protests against the slave trade and its own exclusion from government and public life by the Test Acts. These tensions could be traced back long before the recent events in France to the 1770s, when a small core of dissenters and political reformers had sided openly with the American rebels against the British Crown.

But if the causes of the riots were complex and long-standing, it was the single-minded destruction of Priestley's laboratory that provided their enduring symbol, and it was as 'the Priestley riots' that they passed into history. The spectacle of a frenzied mob dismantling his chemical apparatus gave form to a growing sense that Priestley and his fellow-chemists were playing a dangerous game, one with revolutionary consequences that extended far beyond their laboratories. The symmetry between their chemical researches and their revolutionary politics was unmissable. Their discoveries were trumpeted with claims that that they were shaking loose the very foundations of nature, reducing matter to its previously unseen constituent parts and reassembling them into novel configurations that would transform the world. Nor were the likes of Priestley shy of making explicit the political parallels that they saw in their work. 'This rapid process of knowledge', he had observed the previous year, 'will, I doubt not, be the means, under God, of extirpating all error and prejudice, and of putting an end to all undue and usurped authority, in the business of religion, as well as of science'. In the future that his work envisioned, authority would no longer be able to assert itself simply by force or tradition: it would be obliged to justify itself by reason and consent, 'and the English hierarchy (if there be anything unsound in its constitution) has equal reason even to tremble at an air pump, or an electrical machine'.[5]

For those who took such claims seriously but lacked the chemists' confidence that the consequences of their project would benefit society the Priestley riots reinforced the conviction that the currents of scientific and political revolution were gathering strength from one another, and that the defenders of the *status quo* would need to resort to physical force sooner rather than later. In 1790 the Whig MP Edmund Burke had published his *Reflections on the Revolution in France*, the work that did more than any other to crystallise the reaction against the French Revolution and those who aspired to transfer its ideals to British soil, and that made an explicit case for forms of authority that had never before needed to be defended. In it, Burke had held up to scrutiny the elements in society who believed the next decade would fundamentally remake politics and public life, just as Priestley and his fellow-chemists claimed that they were remaking nature and matter. His most vivid and enduring metaphor for the forces of revolution was drawn directly from chemistry: what the revolution in France had unleashed was a 'wild gas', a seductive but insubstantial vapour of noble-sounding ideals, a smokescreen that obscured the darker forces that calls to revolution were bound to unleash.[6]

Burke's image of a 'wild gas' was a pointed reference to the field in which Priestley excelled: pneumatic chemistry, or the chemistry of gases. Over the previous generation, chemists had shown that the air around them was not simply an inert backdrop to the world of matter, but a complex substance that could be teased apart into a still unknown number of separate elements, many of them with remarkable properties that could never have been guessed at from air itself. In 1767, by capturing the gases that emanated from a brewery close to his previous home in Leeds and bubbling them through water, Priestley had created soda water: it rapidly became a popular health drink, but he had refused to take out a patent on it on the grounds that any benefits from his researches should be free for all mankind. In 1774 he had used his giant lens to release a gas in which candles burned brighter and mice lived longer than in air itself. Since then, his growing collection of glassware and crucibles, evaporating baths and troughs had released ever more mysterious gases from their natural bonds, and with them glimpses of the deep secrets not only of chemical matter, but of respiration and the invisible forces of life.

These artificially produced or 'factitious airs' were powerful propaganda for the chemists' claims that they were able to produce wonders from thin air; but Burke was equally quick to extend the metaphors that they offered in other directions. Those who remade the air we breathe into exotic gases were equally engaged in 'airy speculations' about the direction of society: for all their talk of liberty and fraternity, they were simply addicted to toying with novelty, caring no more for the human objects of their experiments 'than they do mice in an air-pump'.[7] Their experiments carelessly ripped apart the bonds that held the world in harmony: they were not observing nature but torturing it into unnatural forms, not reading the book of life but scrawling over its pages. 'These philosophers are fanatics', Burke insisted; 'they are carried with such a headlong rage towards every desperate trial, that they would sacrifice the whole human race to the slightest of their experiments.'[8] It was no coincidence that their discoveries had destructive and violent applications: Antoine Lavoisier, France's genius of pneumatic chemistry, was also the director of the French Gunpowder and Saltpetre Administration, in charge of providing the explosive substances that, both metaphorically and literally, he and his fellow-chemists were now encouraging an agitated society to light.

The conflict between Priestley and Burke was not simply the age-old struggle between tradition and progress: rather, it represented a rift in British society that had only recently opened and was now widening alarmingly. Not so long before, the two men had been friends: Burke had visited Priestley at Fair Hill to discuss chemistry, Priestley had listened approvingly to Burke defending the rights of dissenters in the House of Commons in 1773, and both had taken their stand on the side of American independence. Burke, indeed, had encouraged the prince of Wales to allow Priestley to dedicate to him his three-volume *magnum opus* of pneumatic chemistry, *Experiments and Observations on Different Kinds of Air*; but the dedication, when it emerged in 1790, was not what Burke had envisaged. Priestley informed the prince of Wales, politely but publicly, that recent scientific discoveries had served 'to expand the human mind', and by the same token to 'show the inconvenience attending all establishments, civil or religious, formed in times of ignorance, and to urge the reformation of them'.[9] For Priestley, the empirical method in science, which put all authority to the test of experiment, naturally extended itself into a test of political tradition against alternatives such as those that had, in 1776 and now 1789, transformed Britain's great neighbours to east and west. An experiment, however, implied by definition an outcome that could not be predicted, and for Burke the established order was far too important to be dissected, exploded, or otherwise tested to destruction.

*   *   *

So it was no coincidence that the Priestley riots had broken out on Bastille Day. The French Revolution was the catalyst for the troubling divisions that Burke and Priestley exemplified, and that were to tear British society apart over the decade to come. The shift in public opinion that was under way was a seismic one, and captured most vividly in the pages of a recently established four-page newssheet, *The Times*. In 1789 the paper, like most of the British reading public, had broadly welcomed the storming of the Bastille, characterising the events in France as a 'struggle between expiring despotism and triumphant freedom', a hopeful sign that France was finally overthrowing the 'despotic sceptre' of Louis XVI and moving towards the kind of constitutional monarchy that had been established in Britain a hundred years previously.[10] But by the first anniversary of Bastille Day in 1790, with much of its French reportage now supplied by traumatised aristocratic émigrés, it was beginning to follow Burke in his suspicions that the violence of the early days of the revolution was the first act in a 'progress through chaos and darkness' that would test British nerves to the limit.[11] Its editorials criticised the incendiary language of the French press and questioned the end point of the 'mad spirit of change' that had taken hold; yet at this point it still accepted that the questions it had raised needed to be debated in Britain too. As chapters of the Revolution Society gathered in provincial towns across the nation, it commented that, 'though we do not agree in opinion with these gentlemen on the principles of the French Revolution', the paper was nevertheless inclined 'to testify our approbation of the motives which produced this zeal in the cause of freedom'.[12]

By Bastille Day of 1791, however, *The Times* had changed its line unrecognisably. The actions of the Revolution Society, it now maintained, 'we have long declared to be highly dangerous'; those who celebrated violence were bringing violent disorder upon themselves. To agitate for revolution was an act 'treasonable to the constitution', and it was 'natural to suppose that some decisive indignation would manifest itself against those factious traitors' who were seeking to 'ripen the lower order of the people into an open aversion to the present system of government'. It was regrettable, but hardly surprising, that 'the loyal spirit of the numerous inhabitants of that great manufacturing town' of Birmingham should have broken with 'uncommon fury' on the fifth column parading in their midst. In *The Times*' account, the magistrates had done everything they could to calm the crowd, but were powerless against their patriotic spirit. Priestley, it claimed, was not what he seemed: he 'who, in public, was the most abstemious man alive' turned out to have 'cellars stored with the choicest liquors'.[13] War was still eighteen months away, and William Pitt was insisting in the House of Commons that ten years of peace with France had never seemed more likely; but *The Times*' lengthy coverage of the Priestley riots in 1791 marked the moment at which the battle lines between revolution and reaction were irrevocably drawn, and the paper's attitude to 'French opinions' was fixed for a generation.

* * *

From Bridgnorth, Priestley travelled south to London, where he took refuge among the long-established dissenting community in Hackney: so intense was the partisan feeling against him that he felt 'I could hardly have been safe in any other place'. His first instinct was to make peace, and he wrote an open letter to the citizens of Birmingham turning the other cheek, and offering 'blessings for curses'.[14] But as the story behind the night's events emerged more clearly, he tempered his forgiveness of the mob with accusations against the establishment who had connived at their rampage. 'I supposed', he wrote bitterly, 'that those who had any influence in the place would be ashamed, or mortified' by the destruction of his property, but 'in this flattering idea' he had plainly been mistaken.[15] The sanctity of property was the cornerstone of established authority, it seemed, unless the property belonged to dissenters. It was clear he could expect no more welcome in Birmingham; even in London, although he felt safe from physical attacks, it would prove impossible for him or his family to make a living. Within two years, and after 'numberless insulting and threatening letters',[16] he would concede that the only future open to him was exile in America.

The Priestley riots were the end not only of Priestley's career in Britain, but of the wider network of the Lunar Society, of which he had latterly been a conspicuous member. He was, in fact, the only member who referred to the society in print. For the others it was a private name for an informal gathering, but for a social and political outsider like Priestley its illustrious company offered a valuable badge of status. The Lunars were, by 1791, giant figures in Birmingham's landscape – the likes of James Watt, Matthew Boulton and Josiah

Wedgwood created wealth and employment to rival the largest landowning grandees – yet the Priestley riots signalled the eclipse of their fellowship. For a generation, their monthly meetings had been an institution, and a prodigious source of the inspiration and invention that had transformed the Midlands and, increasingly, the wider world; but the destruction of Priestley's laboratory gave clear notice that they too might now become fair game for demonstrations of public anger. The Lunar Society met again on 12 September 1791, but subsequent meetings were postponed or cancelled for lack of a quorum.

The Society had, in truth, been in gentle decline for some years: most of its members were ageing, the white heat of their experimental days cooling and the administrative burdens of their enterprises consuming ever more of their energies. But they also suffered from the abrupt change of public mood, and felt a sudden chill towards their endeavours and their influence. Matthew Boulton began to arm the workers in his Soho factory against rioters, and by November 1791 James Watt felt that Birmingham 'is divided into two parties who hate one another mortally':[17] on the one side, the progressive industrialists, the middling classes and the dissenters, and on the other, the traditional authorities of the established church and aristocracy, and the mass of working men under their patronage. It was, as the Priestley riots had shown, a paradoxical conflict, where the authorities were defended by violent mob rule and the revolutionaries pleaded in vain for the forces of law and order to intervene.

It was a conflict for which the rest of the Lunar Society had little appetite. Although they offered personal support to Priestley after the riots, they were unwilling to take up his cause in public: as a spokesman for the dissenting community who saw his ministry as his profession and his chemistry as a hobby, Priestley had always advertised his political opinions more freely than his scientific colleagues. Bastille Day of 1789 had intoxicated them all with hope: Mary Anne Galton, the ten-year-old niece of Erasmus Darwin, recalled later that 'I never saw joy comparable in its vivid intensity and universality to that occasioned by the early promise of the French Revolution' among her family.[18] Darwin, imbibing Antoine Lavoisier's chemistry at the same time as the news from Paris, had confided to James Watt in 1790 that 'I feel myself becoming all French both in chemistry and in politics'.[19] But, especially for the likes of Boulton, Watt and Wedgwood, these idealistic sentiments had to be balanced against the realities and responsibilities of industry and employment. Success in business had made their instincts more paternalist, their focus not on revolution but on rebalancing the rights and responsibilities of the propertied class for the dawning industrial age. War, when it came, would incline them further towards their business interests, increasing the price of iron, placing a vast weight of national responsibility on the machinery of mass production, and offering profits to match.

*   *   *

There was one man, however, for whom the destruction of Priestley's laboratory would be not the end of the story but the beginning.

In the prologue to his *Experiments on Air*, Priestley had made the grandest of claims for the future of pneumatic chemistry. The invisible airs that were beginning to emerge from the laboratory would, he predicted, transform our understanding of nature, and reveal 'principles of more extensive influence than even of gravity itself'. Furthermore, its applications would change the world in ways yet unimagined, and 'of much more consequence to the improvement and happiness of it'.[20] It was a challenge that, after Bastille Day of 1791, was not to be taken up lightly, but it would have one conspicuous champion over the decade to come: Thomas Beddoes, in a fiercely ambitious series of experiments with the newly isolated gases, would steer a course into the eye of the storm. In doing so, he would take the experimental method into uncharted waters, and would prove Priestley's assertion to the prince of Wales that this science had a tendency to 'expand the human mind' by plunging into an unprecedented exploration of the shaping forces behind consciousness itself.

In the course of his researches, the charismatic Beddoes would gather around him a glittering circle of colleagues and friends with whom he would navigate the turbulent decade to come. His experiments would become a crucible where mettle was tested, careers were forged and broken, and from which would emerge some of the great geniuses of the next generation. Their endeavours would blur the boundaries between the laboratory and the wider world, and their project would be swept along in the fast-moving currents of politics and social change around them. Beddoes' experiments would indeed herald a revolution in the improvement and happiness of the human condition, though it would be many decades before the world would be ready for it.

But Priestley had also issued a caveat, which would prove to be just as applicable to Beddoes' experiments as his intimations of future glory. 'Nothing is more common' in experiment, he noted, than 'the most unexpected revolutions of good or bad success'. Even pneumatic chemistry, in which the secrets of nature lay hidden in plain sight, might not be immune from receiving 'a check, even in the most rapid and promising state of its growth'.[21] Any experiment worth undertaking must, as Burke insisted, offer the possibility of failure, and it was a possibility to which Beddoes exposed himself with great courage, and at times great recklessness. As he drove his experiments forward in the teeth of bitter opposition, he found himself tacking between triumph and disaster, tragedy and farce, his experiments becoming both legendary and notorious. A wild gas, as Burke put it, had 'plainly broken loose', but it would be a long time before the experimenters would be able to see 'deeper than the agitation of a troubled and frothy surface', and to understand precisely what had been gained in the course of their researches, and what had been lost.[22]

Part 1

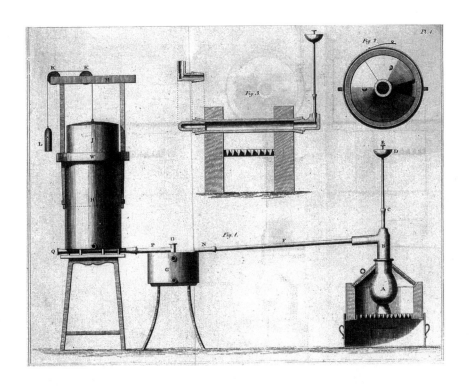

1

Freedom's Garland

Birmingham still resembled a battlefield when Thomas Beddoes passed through it the following weekend. Although the riots had subsided with the arrival of the militia, the streets were still littered with charred beams and broken glass, and the walls covered with graffiti damning Priestley and all his works. It was a spectacle that made a profound impression on Beddoes, and its significance for him would deepen as the decade progressed. Bastille Day of 1789 had seemed to him to be a moment when the world had finally caught up with his own dreams and ideals, and aligned its direction of travel with his; but Bastille Day of 1791, he would come to realise, was an equally defining moment around which the course of his life would need to be reconfigured.

Beddoes was a regular traveller through the city. His home town, Shifnal in Shropshire, was linked to it by a modern and relatively speedy turnpike road serviced by post-chaise coaches; from it, another busy carriage route took him to his residence in Oxford, where he held the post of chemical reader at the university, and from where he was now returning home for the long summer vacation. Birmingham was hardly a scenic stopover at the best of times: over the previous generation, it had evolved into an urban landscape profoundly alien to most English people, a foretaste and warning of the century to come. To visitors, it felt less like a town than a giant factory, or, as Edmund Burke called it, 'the grand toy-shop of Europe'.[1] A thousand small workshops and production lines churned out toys and buckles, cutlery, buttons and snuffboxes, flooding the country, and increasingly the world, with smartly designed trinkets at rock-bottom prices. It was famous above all else for Soho, the vast steam-engine manufactory and self-contained workers' town opened in 1762 by the Lunar Society's Matthew Boulton and James Watt; but the majority of Birmingham's army of craftsmen worked not under Boulton and Watt's benevolent and progressive paternalism, but in conditions that beggared the eighteenth-century imagination. The clank and din of wheels, furnaces and forges never ceased; the air was thick with the acrid and oily scent of burning, grinding and polishing, and the city was dusted with a fine layering of soot, tar and clinker.

The impact of this level of industrial production on the city's inhabitants was, to many observers, terrifying. Robert Southey, passing through a decade later, recorded that he was unable to comment on the health of the population, 'having seen no other complexion in the place than what is composed of oil and dust smoke-dried. Every man whom I meet stinks of train-oil and emery.'[2] Bloodshot eyes peered out of blackened faces, those employed in the brass and copper works frequently topped with green hair. Whole families of occupational diseases, particularly chest and lung conditions from inhaling noxious fumes, were becoming endemic. A common feature of the reportage on the riots had been lurid descriptions of the physical specimens who had emerged from the warrens and factories to be caught in the spotlight of the nation's attention: the 'bunting, beggarly, brass-making, brazen-faced, brazen-hearted, blackguard, bustling, booby Birmingham mob'.[3] One local shopkeeper, having had his property ransacked, had been dragged to a dingy tavern where he had been 'forced to shake a hundred hard and black hands' and buy 329 gallons of beer before the dirt-encrusted rioters allowed him to leave.[4]

Yet in many ways more troubling than Birmingham itself was the question of what might be done to ameliorate it, and what would happen if the rest of the country began to follow its example. The riots had exposed a community that was ungoverned and essentially lawless: forgery of coins was becoming as big a business as the making of toys, and it was said that any currency with which Britain traded, in Europe, India or America, circulated freely in Birmingham in counterfeit form. Guns were now being manufactured in vast quantities for overseas markets such as Africa, mostly without any effective regulation or control. There was no Member of Parliament to represent the interests of Birmingham's citizens, since the constituency system dated back to the time when it had been no more than a small cluster of villages. Nor did the mob show any interest in participating in the national politics: the riots had been sparked not by their hunger for parliamentary reform, but by their hatred and resentment of those who were calling for it on their behalf.

*   *   *

Thomas Beddoes made a striking first impression. At the age of thirty-one he was, if not yet a public figure, at least a man easy to recognise and hard to forget. On at least one occasion over the summer of 1791, his reputation preceded him: passing through Bristol, he found himself being discussed in his presence by his fellow coach passengers, clearly unaware that he was among them. 'Dr. Beddoes of Oxford', he learned, is 'so fat and short he might almost do for a show'. Yet his opinions were held to be even more freakish than his appearance: 'excepting what he might know about fossils and such out of the way things', his fellow-traveller continued, 'he is perfectly stupid and incurably heterodox'.[5] Beddoes quietly encouraged the conversation until he remembered that he was holding a book on basalt formations in his hand, and indeed that his fellow-passengers had just been admiring the plates; he attempted to melt into the background, and escaped without being identified. Short, rotund and wheezing he certainly was, if not quite the physical grotesque that these strangers had been led to expect; but he was also a restless ball of energy. He would spend the summer walking cliff paths and scaling granite tors in fierce gales, and by the time he returned to Shropshire in the autumn he would be fit enough to walk twenty miles without stopping, and with a pile of books under each arm.

To those who knew him, however, Beddoes' most striking quality was the staggering breadth and depth of his learning. As chemistry reader at Oxford's Pembroke College, he was one of only a tiny handful of men in Britain to make a professional living from the science: he lectured to packed halls, boldly systematising the torrent of research that was flowing into the country from France, Germany and beyond, his knowledge rapidly superseding that of the pioneering generation who had taught him. Of pneumatic chemistry, one of his early specialisms, he had long known all that there currently was to know, and indeed had recently reintroduced a crucial body of knowledge that had lain forgotten for a century. In the university laboratory, he was currently experimenting with air pumps and barometers and, just recently, had become a pioneer in the nascent technology of hot-air ballooning. In June 1790 he had caused a sensation by collaborating with James Sadler, the first Englishman to make a manned balloon ascent, on the construction of an unmanned sphere filled with inflammable airs and lit by touchpaper, which had ascended in a fiery arc over Oxford's spires. It was, by Beddoes' own account, 'one of the most perfect and beautiful experiments I have ever seen'.[6] He had never been lucky enough to see an igneous meteor fall, but he was satisfied that he had created, in some respects, a fair facsimile.

Yet if Beddoes had known no chemistry at all, he would still have been a prodigious polymath. In addition to Latin and Greek, both of which he read habitually, he had also taught himself French, German, Spanish and Italian; he had translated scholarly works from all these languages and was well versed in their science and literature. He was a keen and systematic botanist who had written an entire *Flora Britannica*. He had an eidetic, perhaps even photographic memory, and an endless store of arcane knowledge; he was a formidable whist player, with a reputation for being able to remember the turn of every card. His German, a rare accomplishment among British scholars, gave him access to the new biblical criticism, and he was familiar with the latest research not only on the history of the Holy Land but on other ancient civilisations, such as the Vedic and Brahminic cultures of ancient India. He was closely involved with the leading experimentalists of the industrial revolution, and was currently composing a paper on the differing properties of cast and malleable iron and their basis in chemical impurities. He was an omnivorous reader of literature and an accomplished poet. Yet more urgent to him than all of these was the subject that had been his ruling passion from an early age – medicine – and the one with which he was at present most intensely engaged, which was geology.

Despite all these accomplishments, and excepting that he usually carried a pile of books with him, there was little of the typical scholar in Beddoes' engagement with the world. He could be stern with his juniors and uncomfortable with formality, but more often he was bustling, bubbling and sociable, and anxious for the happiness of those around him. His favourite novel was *Tristram Shandy*, which he valued particularly for its indulgence of eccentricity, its tolerance of paradox and the way it 'put people in good humour with the follies of one another';[7] and he himself was rich in Shandean qualities, constantly in motion and with an opinion on everything, leaping from topic to topic with cranky scholarship, puckish wit and a genial disregard for linearity. His was, as his companion of the summer of 1791 would later describe it, a 'congenial muddling disposition': any subject might unpredictably jump tracks to a second and a third, vast tapestries of thought were constantly under construction and ambitious schemes for future projects added to an ever expanding list.[8]

But as he set out in the summer of 1791, Beddoes was beginning to feel the need to pick from his array of accomplishments a specialism that he could make his own. The discipline he would choose would be one that did not yet exist, but that would come to define the rest of his life.

*   *   *

He was travelling back home from Oxford to meet one of his students, a tall, fair young man named Davies Giddy, with whom he stood at the beginning of a warm friendship that was to last a lifetime. Giddy presented a sharp contrast to his teacher not just physically, but in almost every respect. Where Beddoes bubbled over with projects, theories and experiments, and shared them with all and sundry, Giddy was grave and reserved, keeping his own counsel with a combination of adolescent shyness and maturity beyond his years. The virtues of caution and the perils of recklessness had been instilled in him from an early age, a process that showed no sign of relaxing: his father had accompanied him to Pembroke College and remained there to chaperon him, fearful of his exposure to the city's temptations, despite the master's opinion that the eighteen-year-old was already 'sufficiently prudent to remain at Oxford alone like other young men'.[9] This protectiveness had made Giddy easy with the company of his elders, but had also fed a diffidence that would persist in his relations with his contemporaries and with women in particular, with whom he scrupulously engaged on a platonic and brotherly basis, whatever the evidence on either side of other forms of desire.

Now aged twenty-five, Giddy had finally slipped his father's attentions to pass the early summer unaccompanied in London, much of it spent either running errands for Beddoes or taking advantage of his introductions to the worlds of science and politics. He had been seeking out dealers and buying samples of amber and rare minerals, and overseeing the printing of Beddoes' current paper, and had taken up introductions to visit distinguished figures such as Sir Joseph Banks, president of the Royal Society – a post that Giddy, in a future at this point unimaginable, would eventually hold himself. He had also taken the opportunity to gain first-hand intelligence of the day's great political debates on which, under Beddoes' wing, he stood enthusiastically on the side of revolution and reform. He had met with Tom Paine and Horne Tooke, the waspish veteran of the old Wilkes and Liberty agitations, and had visited the House of Commons to watch one of Edmund Burke's marathon exchanges with Charles James Fox. He was now making his way up to Beddoes' family home in Shifnal, wearing a French tricolour cockade in his hat and carrying a letter of introduction to Joseph Priestley. He had made it as far as Coventry by Bastille Day, when the news from Birmingham forced him to change plans.

Beddoes, following on a couple of days behind, hurried through the smouldering aftermath of the riots and caught up with Giddy at his family home in Shifnal, where he proceeded to conduct him on a week's tour of Shropshire and its industrial marvels. Here, the forces that had transformed Birmingham into a soot-blackened vision of hell were also vigorously at work, but had been contained in pleasing harmony with the rolling, fertile landscape and its traditional forms of rural life. Small-scale miners and colliers, who had operated on common land for generations, were expanding their activities with steam water-pumping engines and iron rails to carry skips filled with coal and ore; Beddoes and Giddy inspected the smelting and coating furnaces that now squatted among farms and forested hills. Twice they visited Coalbrookdale, the largest and most technically advanced of the county's iron works, where relatively poor iron ore was reduced with surpassing effectiveness by means of fluxes derived from the local limestone, and where the dark, peaty waters of the river Severn had, since 1779, flowed beneath the spectacular arch of the world's first iron bridge, its 375 tons of cast metal braced between the steep and ruggedly forested banks.

There were many parallels with Giddy's home county of Cornwall, which was to be their destination for the rest of the summer. Here, too, traditional mining was being rapidly industrialised, and engine houses, smelting works and chimneys studded the desolate expanses of heath and moorland. These were developments in which Giddy would soon be playing an important role, but the immediate purpose of their stay in Cornwall was to pursue more fundamental scientific enquiries into the nature of the earth itself. This was the subject on which Beddoes had recently delivered a series of exceptionally well-attended lectures at Oxford, and in which Giddy had graduated from his student to his apprentice, factotum and sounding board.

Geology had been among Beddoes' consuming interests since his enrolment at Edinburgh University in 1784, when he had studied chemistry under the brilliant tutelage of Joseph Black, with whom he still maintained a vigorous correspondence. Black had revolutionised chemistry in the 1750s with his work on gases: by isolating 'fixed air' – carbon dioxide – he had demonstrated not only that it was the same substance as the miners' suffocating 'choke-damp' but that it played a hitherto unsuspected role in many chemical reactions. He had proceeded to isolate many more gases, or 'airs', and to apply chemistry to the new demands of industry in ingenious ways: he discovered that bleaching agents could be made cheaply from kelp, and that ships' hulls could be protected from woodworm by coating them with tar. But Black was also keenly interested in the parallels between the effects he could generate by heat and pressure in his laboratory and the formations of rocks and minerals in the earth itself. His lectures began to include speculations about the formation of strata and the effects of high temperatures on quartzes and granites, and Beddoes had been inspired by them to spend the summer of 1785 travelling rough across the wilderness of the Scottish highlands searching for traces of volcanic formation among the crags and glens.

By the late eighteenth century there was much evidence to suggest that the age of the earth far exceeded the limits of biblical chronology. To the famous conundrums of Classical and Renaissance philosophers – the presence, for example, of fossilised seashells on mountaintops – had been added much study of the formation of strata, and observation of the gradual processes of erosion and deposition. These discoveries had been squared with the biblical account by the prevailing orthodoxy which held that the newly created earth had been covered by a universal ocean, identified by some with the biblical Flood, that had exposed and weathered the continents as it had receded. But Black, and Beddoes with him, had grasped that the world was not crumbling away from a state of original perfection, but was constantly replenishing its lost mass in a self-renewing and homeostatic system that had been operating through a previously unimagined immensity of time.

A friend of Black's, James Hutton, a crofting farmer and independent scholar, had first laid out the foundations of this startling theory at the Royal Society of Edinburgh in 1785. After years of observing rock strata and calculating the timescales required for them to form, buckle, erode and surmount one another, Hutton's conclusive moment of discovery had come in 1788 at Siccar Point on the east coast of Scotland, where he 'grew giddy from peering into the abyss of time' as he realised that he was looking at rocks that had been laid down over millennia on the sea bed, then had buckled upwards to form part of a massive mountain range, then had been ground down particle by particle to sea level once more, and been covered once more, grain by grain, with new sedimentary strata. It was a deduction that implied an endless process on a cosmic scale, with 'no vestige of a beginning, no prospect of an end'.[10] While Hutton still regarded structure and strata as the crucial evidence for this process, Black and Beddoes had been quick to recognise the potential of chemistry to demonstrate that particular types of rock, such as granites and basalts, had been created by extremes of heat and pressure deep beneath the earth's crust.

To most contemporary geologists, these were interesting but moot speculations, and would remain so until their proposers were able to move from analogy and supposition, and the postulation of implausibly vast forces and timescales, to direct evidence and proof. Beddoes, however, was convinced that such proof was available. He had spent many hours in the furnaces at Coalbrookdale, watching the smelting of iron, the heaving of the molten mass of ore, the comminution of the rock to gravel, then to fine sand that formed into crystal as it cooled. He had observed that rapid cooling produced a glassy mass, but that the crystal lattices in the structure became more marked when the mass cooled more slowly. This presented a spectrum that was closely matched by his collection of basalt samples, which showed a similar range from glassy uniformity to nests of hexagonal needles, and a paper that he had written to this effect had been read before the Royal Society in London in January.[11]

It was a finding that, if he could substantiate it, would present a significant challenge to the prevailing view that the world was shaped simply by weathering and receding waters. Those who held to this theory, of whom the most authoritative was the German mineralogist Abraham Gottlob Werner, did not deny the existence of volcanoes, or the lava and pumice that flowed from them, but they held these igneous materials to be localised and distinct from the rocks that made up the bulk of the planet. Their contention was that crystalline basalts were not volcanic but aqueous, their crystals precipitated from oceanic salts, and that this explained why they graded into other aqueous forms such as granite, porphyry and serpentine. Beddoes, however, believed he could demonstrate that all these rocks were volcanic, and that the distinction between vitreous and crystalline forms, held by Werner to be a defining one, was in fact a greyscale that reflected their origins as varieties of lava, extruded in molten form from the pressurised depths of the earth and crystallising as they cooled.

There were several reasons why Cornwall was the ideal place to comb for the evidence Beddoes sought. It was one of the few extremities of Britain that Hutton, in his years of assiduous geologising, had never visited; the means were now at his disposal, and with them the prospect of an agreeable summer with Giddy. Above all, its rugged terrain was a geologist's paradise. Cornwall's moorlands were dotted with spectacular tors of weatherbeaten granite; its cliffs exposed dramatic strata shot through with faults, thrown up at crazy angles, in some cases even turned upside down. Its abundant tin and copper deposits meant that it had been mined more extensively and for longer than anywhere else in Britain, and its inhabitants had long been familiar with the exotic forms of volcanic rock – greenstone, dolerite, gabbro, elvan – that intruded through the slate bedrock in veins known as dykes or sills. The range and complexity of these volcanic forms was immense, and Beddoes was already beginning to suspect that they indicated a vast lava pillow on which 'the whole western side of our island has probably been raised up'.[12]

*   *   *

By the end of July, Beddoes and Giddy had reached the fringes of Devon, and left the turnpikes for the notoriously poor Cornish roads: most no more than bridleways, and coaches a rarity among the daily traffic of packhorses. A further week's rough progress brought them to Tredrea, Giddy's family home at the far south-western tip of Britain. A few miles beyond it lay the sparkling expanse of Mount's Bay and the busy port, customs post, market, mining and tin-stamping town of Penzance; but Tredrea, inland, was surrounded by sparse, stony fields and low hills of gorse and scrub, and linked only by muddy tracks to the huddle of squat granite buildings that formed the local village, St Erth. Beddoes' first impression was of an outpost of monotonous and 'dreary desolation'.[13]

The house at Tredrea partook of the gloom of its surroundings. Giddy's grandmother had inherited it along with its surrounding orchards and fields, but since there was no other form of income, family and guests alike had shivered around small fires that failed to drive the damp from the thick granite walls, and stumbled through cavernous darkness for want of money for candles. But Giddy's father had, by diligence and hard work, qualified to study divinity at Oxford and become a vicar, and the family had subsequently recovered the status of respectable minor gentry. Like the landscape, Beddoes found the first sight of the house forbidding, but was swiftly warmed by the close and affectionate ambience in which Giddy lived with his parents and younger sister, Philippa.

Both Giddy and Beddoes would always recall the long summer of 1791 as a golden season. They made energetic excursions around the coastal paths and beaches of the Land's End peninsula, exhilarated by the buffeting winds and the wildness of the rocky shores. Far from the cities and their Church and King mobs, they wore their French cockades with pride, openly declaring themselves part of the new brotherhood of man. At a Cornish Copper Company dance, Beddoes even took the liberty of removing his tricolour and pinning it to his partner's hair, breaking into verse to hail the sight of 'Freedom's garland on Beauty's radiant brow!'[14] They tramped from north to south coast through the bracken and stunted hawthorns of the high moorland, where the sea formed the horizon beyond and behind them, and curious stone circles and menhirs, attributed by the local antiquarian William Borlase to the ancient druids, rose out of the sea of gorse and heather to puncture the skyline. They scrambled up and down the exposed vein of serpentine, polished to a sea-green marble by fishermen's boots, above the nested white sand bays of Kynance Cove, which Beddoes would remember for the rest of his life as 'that charming habitation of the sea-nymphs'.[15] They travelled rough and spontaneously, sleeping in cots and on truckle beds in whatever village or harbour they found themselves at dusk. Beddoes deferred his departure until the end of September, and then stayed on through October.

As they walked, they talked endlessly of politics, and in particular of the news from France. It was already clear that this was the great experiment whose result would determine the future of politics in Britain and beyond. 'How much will the National Assembly have to answer for', Beddoes had observed earlier in the year, 'not to France only but to mankind, if they should not establish a good and prosperous government'.[16] The fledgling Assembly had already made bold progress, abolishing hereditary titles and seeking to bring the vast power and wealth of the Catholic church under the wing of elected government; but it was equally clear that obstacles and pitfalls were seething like snakes around its cradle. Louis XVI's flight to Varennes in June, and the manifesto he had left behind cheerfully confirming his perfidy and contempt for the constitution, had strengthened the calls not merely to set the monarchy within a consensual frame, but to eliminate it entirely by establishing a republic. As these previously marginalised voices grew bolder, the Assembly edged precariously towards martial law: in July, a crowd supporting the new republican petition had been fired on in cold blood by the army at the Champ de Mars. On Bastille Day, as Birmingham burned, a night of bloody violence in the swamps of the Bois Caiman had announced the rebellion of the slaves in Santo Domingo: by August, while French royalist and republican factions jostled for control of the island, the Assembly was being forced to choose between extending their principles of liberty to the Caribbean colonies and keeping hold of their greatest source of wealth. 'One vexatious event treading in this manner upon the heels of another', Beddoes worried, 'this must prove by too severe a trial the infant strength of the constitution! & what if the Assembly should prove unequal to their functions?'[17] The French people, never having experienced political justice, were now expected to administer it perfectly at the first attempt.

In Britain, meanwhile, the backwaters where young men could wear tricolour cockades and make public salutes to liberty were rapidly disappearing. Beddoes and Giddy saw the French cause as a moderate and rational one, a search for a settlement agreeable to all estates, and by the same token similar if not identical to that they wished for in Britain herself. 'I dread the Paris mob',[18] Beddoes had written to Giddy before the summer had begun, and it was a sentiment in which he never wavered: the anarchy threatened by mob violence would remain for him at least as great a danger as the resurgence of the *ancien régime*. But, on both sides of the Channel, this middle ground was eroding beneath its proponents' feet. Edmund Burke's predictions for the French Revolution, ridiculed by many on their publication for their alarmism, were proving horribly accurate; and the response to them that Tom Paine had published in March, under the title *Rights of Man, Part I*, had driven the counter-argument to ruthless conclusions that moderate reformers found extreme and divisive. Even Giddy, on his visit to the House of Commons, had found himself half-persuaded by Burke. Beddoes was concerned that 'an able and popular answer to Burke'[19] was sorely needed, but he believed that Paine's levelling republicanism was a prescription for anarchy. He was pinning his hopes on the work-in-progress of an old Edinburgh acquaintance, James Mackintosh: where Paine rejected Burke's entire architecture of political authority, Mackintosh engaged respectfully with it, and attempted to rehabilitate French events within it. When Mackintosh's work emerged under the title of *Vindiciae Gallicae* it impressed many, including Burke himself; but its delicately finessed middle ground, like most moderate positions on the revolution, would be swiftly forgotten. Even Mackintosh himself would soon reject it, and take his place alongside Burke among the revolution's sworn enemies.

Meanwhile, Beddoes and Giddy's geological researches were producing an abundance of rock specimens showing intermediate forms between granites and basalts, and variations in crystalline structures between coarse and fine granulation. In the cliffs around the harbour of Porthleven, they found clear evidence that the dykes, as the Cornish called them, were veins of volcanic rock that must have shot through the older layers of slate in molten jets under high pressure, from which it was logical to conclude that they represented 'a large mass of similar matter in the depths of the earth'.[20] Beddoes was assembling in his mind the modern geological narrative that Cornwall's slates had originally formed on the sea bed before being uplifted by the pressure of molten lava that had burst through them and cooled to form granite, raising the beds above sea level and precipitating chemical deposits of crystal and metal in the process. But when the excitement of the chase had faded, Beddoes would be candid in his conclusion that their discoveries amounted to something less than proof. 'I learned in Cornwall', he wrote later to Giddy, 'that the moment you came to particulars, that moment you feel a most humiliating sense of the helplessness of your philosophy'.[21] It was one thing to form a theory and gather evidence to support it, but quite another to demonstrate that this evidence could point in one direction only.

* * *

But Beddoes and Giddy had another preoccupation, besides geology and politics, that would develop into their summer's most enduring theme, and signpost the route that Beddoes' career was poised to travel. In his wanderings around Cornwall, Beddoes had, as was his ingrained habit, surveyed the state of health and sickness around him, and had been struck by the prevalence of occupational diseases such as miner's lung which, combined with the omnipresent consumption, meant that the county had a large population of chronically sick in need of treatment. Cornwall had no general hospital for treating the poor, but over the summer a local Whig MP, Sir Francis Bassett, was campaigning for a public infirmary, funded by the county, in the fast-growing stannary town of Truro. Beddoes, however, was of the opinion that it would be more effective to establish a network of local dispensaries, and in the second half of August he and Giddy sat down to make their case, which they published as a pamphlet in September.

Of all the roles that Beddoes inhabited, that of doctor was perhaps the most deeply rooted, and the closest to a unifying theme in his life. The primacy of medicine, he would argue on many occasions, was a simple utilitarian calculus: nothing else was so directly effective in increasing the sum of human happiness and alleviating pain and misery; and what else could be the ultimate purpose of learning? Yet his engagement with it was far more intimate than such dispassionate logic would suggest. After his death, Beddoes' first biographer, John Stock, would unearth a story from his Shropshire childhood to stand as an origin myth for his medical calling; and despite its psychological neatness, there is no reason to doubt its truth. Its source, Dr William Yonge, the family doctor in Shifnal, remained a lifelong friend of Beddoes, and indeed would become an early contributor to the project with which he would attempt to change the face of medicine.

Thomas Beddoes had been named after his grandfather, the man who had raised the family from farming to gentlemanly stock. The latter was a 'man of strong mental powers and of great personal activity' who had built up a tanning business, acquiring land as he went, until he had 'by his industry and enterprise acquired a considerable fortune'.[22] The young Thomas idolised his grandfather, who awoke him to the industrial applications of chemistry at an early age by showing him how to treat and dye leather hides. But when Thomas was nine, his grandfather suffered a tragic accident. He was thrown from his horse, just outside his front door, onto a pile of timber; his ribs were broken, and one of them pierced his lung. The punctured cavity collapsed, and emphysema took hold; the condition was grim, and terminal. He was taken to his apartment and attended by a series of doctors, his features horribly discoloured and distended as he struggled to fill his bubbling and collapsing lungs.

Throughout this fatal crisis, young Thomas sat quietly by his grandfather's side, watching the doctors' ministrations gravely and intently. Dr Yonge, who was among them, began to notice in the boy 'instances of extraordinary acuteness', and to answer his questions and encourage his interest in anatomy, disease and medicine. After his grandfather's death, Thomas got into the habit of spending 'a great proportion of his leisure hours in the shop and surgery of his new friend'. He learned to use pestle and mortar, and to make herbal and pharmaceutical preparations; the adults around the house began to refer to him as 'the little doctor', a nickname that gave him great satisfaction. From this point on, 'whenever interrogated upon the subject of his future profession, he uniformly replied that he would be a physician'.[23]

This was indeed the course that he had followed. He had received his MD from Oxford, and with it qualification to the Royal College of Physicians, and had then progressed to the advanced teaching at Edinburgh University, where the old scholastic tradition of Galenic and humoral theory had been superseded by a vigorous programme of anatomy and experiment, and the search for a new understanding of the physical basis of life and the elusive forces that animated it. He had studied there with distinction, and by the end of his three years had become president of the university's Royal Medical Society. It was his pursuit of the most advanced knowledge that medicine had to offer that had originally brought him under the thrall of his professor, Joseph Black. Black, although he was a physician, rarely practised; his great expertise was in chemistry, and his lectures on the subject were, Beddoes declared, 'the best I have ever heard or ever shall hear'.[24] Under his tutelage Beddoes had become engaged in the great debates that were on the cusp of transforming chemistry into the fundamental language of nature, and it was essentially Black's programme that Beddoes was now continuing and extending with his readership at Oxford.

He had, nevertheless, continued to practise medicine, not as a career so much as an extension of his personality. His progress through the world was accompanied by a running commentary on sickness and health: his letters to Giddy typically included bulletins on the medical state of the nation ('I find that contagion has discharged a full blast of the pestilential breath over this country. Putrid fevers and sore throats abound').[25] His retentive memory had made him a walking storehouse of remedies that he had gleaned from other doctors or witnessed at first hand, and which he was constantly putting to the test. Whether from a simple imperative to do good, or a more complex need to do for the world what the doctors had been unable to do for his grandfather, Beddoes would always be a doctor, whatever other profession he might avow.

Yet although he identified himself as a physician, his relationship to the rest of the medical profession was an abrasive one. He was not its ambassador so much as its scourge and its corrector: medicine, in his view, was in need of a drastic reformation. The discoveries that had transformed the sciences over the previous two centuries had left it largely untouched, still awaiting the transformative genius of a Boyle or a Newton. Even the handful of great advances, such as William Harvey's discovery of the circulation of the blood, published in 1628, had failed to generate new forms of treatment. Physicians still tended to see their role in Hippocratic terms as the providers of a gentle and holistic framework of care, intervening as minimally as possible to restore the balance of health, with chemical fixes regarded as a risky and final resort. Their treatments consisted predominantly of herbal palliatives, dietary recommendations and lifestyle advice; and since these were also on offer from self-appointed healers, nurses, midwives, clergymen and cheap home-medicine almanacs, it was unsurprising that qualified physicians with their expensive services were vastly outnumbered by herbalists, quacks and a vernacular culture of medical advice. Operating in a marketplace where all competed for the minority of wealthy patients, most doctors were more concerned to protect their trade by husbanding their few secrets than to alleviate suffering by spreading them among the wider population.

The pamphlet Beddoes now produced with Giddy, entitled *Considerations on Infirmaries, and on the Advantage of Such an Establishment for the County of Cornwall*, was his first salvo in what would become, over the next decade, a unique public conversation about the role and future of medicine, and it contained several of the prescriptions that he would continue to insist were essential for the reformation of the profession. He regarded a large central infirmary, of the kind Sir Francis Bassett was proposing, as a prime example of the physician's tendency to elevate his own convenience over that of his patients. Practitioners naturally preferred to concentrate their resources and demand that patients come to them, but many of the poor and rural sick were in no position to do so: central dispensaries would always receive a self-selecting sample that would exclude those in greatest need. Rather than building a temple of medicine for patients to visit in supplication, medicine should seek 'to equalise and diffuse medical skill, which it can never be in the interest of so large a county to confine to one spot', and to bring itself as close as possible to those who were in most need of it. 'Dispensaries might easily be established in every town', Beddoes and Giddy suggested, through which medical resources could be distributed 'with much greater convenience to patients'.[26]

Yet even in a short handbill, Beddoes was eager to expand the horizons of the debate, and to demonstrate that medicine was not simply the practice of an elite professional group but a thread that needed to be woven deep into the texture of public life. Neither infirmary nor dispensary, he concluded, would make half as much difference as making the poor aware of the benefits to their health of 'having the windows of their habitations enlarged and better disposed for the renewal of air, one great source of epidemic diseases'. 'This defect', Beddoes had noted on his travels over the summer, 'seems to prevail nowhere more than in Cornwall'.[27] The local architecture of thick-walled and small-windowed granite and slate dwellings had prevailed for centuries, a traditional response to living in the hanging mist of high moors or being whipped by salt spray from Atlantic gales; but its effect was to leave the majority of the population chronically choked and poisoned by the smoke from their own hearths. If medicine was to develop its potential as fully as Beddoes envisaged, apparently immutable features of the human landscape would need to be reconfigured.

*   *   *

As Beddoes pondered with Giddy the problems of Cornwall's health and the choice between general infirmary and local dispensary, he began to conceive of an initiative far more ambitious than either. Later, he would present his scheme as an epiphany that arrived suddenly, unexpectedly and fully formed, and would insist that he could 'almost remember the spot in the field' between Tredrea and St Erth where it had struck him.[28] In truth, parts of the idea must have been with him for some time, and others would only slot into place at a later date; but it was in the muddy cattle pastures of Tredrea that he was struck by the possibility that the next step in his career might be one that combined and distilled all his scientific and medical learning into a project that would benefit humanity on a scale never previously conceived.

In a sense his scheme flowed logically from the calculus of benefits that he cited as his rationale for doctoring: just as medicine offered the most underexplored and potentially beneficial application of science, so the most underexplored and potentially beneficial application of medicine was the treatment of consumption. As Beddoes observed in his pamphlet, it was tubercular, occupational and other chronic disorders of the lungs that weighed most heavily on the Cornish people; and the Cornish were no exception. Consumption, also known as phthisis but yet to be known as tuberculosis, was a disease, or complex of diseases, of obscure origin and terrifying progress. Like cancer, it was a protean form of death itself that consumed the body from the inside out, starting almost anywhere. In his pioneering estimates of its national prevalence (among which other chronic lung conditions such as emphysema were likely included), Beddoes would reckon that one in every four deaths in Britain were due to *Phthisis pulmonaris*, and that once it had taken a firm grip on a subject it was almost invariably fatal.

Consumption was a condition that was poorly understood and, then as now, popularly associated with delicate and romantic sensibilities; but Beddoes had seen enough of it to know better. 'Writers of fictitious biography', he would later observe, may depict it as 'a state on which the fancy can agreeably repose', an image of 'a blossom nipped by untimely frost';[29] but for Beddoes, 'the oftener I see it, with the greater horror I do view it'. In its early phases, the symptoms might amount to no more than a light cough, with no accompanying pain; in postmortem dissection, its only signs might be white spots on the tubercules of the lungs; but from these slight indicators could flow a procession of agonies unmatched by any other condition. In the absence of a cure, one of the most effective tools of medicine was exact observation and description of disease, and Beddoes' descriptions of consumption, as of many conditions, were unexcelled in the medical literature of the day. It could bring the 'harrowing chills and scorching heats of fevers', the 'drenching and draining sweats of ague', the 'oppression of asthma' and the 'insupportable languor' of slow wasting diseases. As it progressed, it could produce an 'actual drowning by inches', as the chest filled with fluids that the patient had no longer strength enough to clear; and by its end, as the breath was slowly strangled, it could progress to an unendurable pitch of 'agony that will last for hours, even for days'.

'If you have a feeling heart', Beddoes concludes, 'you must wish me to put an end to my description . . . I will therefore stay only an instant while I observe that the more shocked you are the better'.[30] He would never stop repeating the litany of consumption's hideous consequences – 'the perpetual pestilence of our island'[31] – and the disease would take on something of the aspect of a personal demon. It was a curse of which he was reminded daily by his physician's duties, and more intimately still by his own wheezing chest, precisely the type of weakness that the disease waited patiently to exploit.

The business of treating consumption was a major preoccupation of eighteenth-century medicine and, typically, one massively skewed towards the wealthy. It had long been observed that some patients, particularly in the early stages, could win remission from it by a change of air: sea breezes, dry air and sunshine were all sought out in seaside and spa resorts or extended Alpine and Mediterranean travels. For those with the leisure and the resources, there was also undoubted benefit to be had from bed-rest, exercise programmes and special diets. All these, however, were palliatives that might ease the symptoms but that left the root cause untouched; the doctor's skill was limited to painstaking administration of a regime of comfort and exercise, airs and purgatives, and interpretation of the obscure signatures of sweats and stools, blood and breath. Claims for an actual cure for the condition were limited to the quacks, empirics and faith healers whose clientele were the desperate, in the final stages of the disease.

For Beddoes, this state of affairs exposed the rotten heart of medicine. The doctors' palliative care was just effective enough to make it desirable, and to sustain a pricing structure that produced a fat living for those who could sell their services to the wealthy. But consumption was a contagious disease, and the untreated poor offered it a permanent reservoir; he could not understand why physicians were not 'tormenting themselves with perpetually reiterated efforts to subdue so terrible an enemy to the human species'. 'Can anyone doubt', he asked, 'whether it be criminal in any practitioner of physic, whose imagination can suggest a new plan, in the smallest degree plausible, for the treatment of consumption, not to pursue it?'[32]

The project that Beddoes conceived in the fields of Tredrea was to develop a new route to conquering consumption by the application of pneumatic chemistry. If changes of air were effective in treating consumption, might not the new science's artificial or 'factitious airs' hold the key to a vastly more effective form of therapy? Consumption, as a disease of the lungs, had thus far been treated chemically with pills, ointment or salves, which by their nature could only address the symptoms of the condition; but airs could be administered directly to the seat of the disease and its primary manifestation, the lesions in the lung. Beddoes would bring the latest discoveries in pneumatic chemistry into the sphere of medicine by setting up a combined pneumatic laboratory and infirmary where he would isolate all the airs that chemistry could produce, and systematically test them on patients suffering from consumption and other lung conditions.

It was an ambitious idea; in fact, it was several enormously ambitious ideas combined. It was striking enough to propose that the invisible airs coaxed from laboratory troughs and air pumps might constitute a new family of medicines; but it also implied a bridge between two institutions, the laboratory and the hospital, that had thus far been entirely separate. It would be, as he later put it, 'the first example, since the origin of civil society, of an extensive scheme of pure scientific medical investigation'.[33] New chemical discoveries in medicine emerged only rarely, and usually by trial and error, followed by a gradual osmosis from one practitioner or region into the mainstream. Beddoes' vision of a formal institution for medical research implied not only a new conception of chemistry but an equally new conception of medicine, where sick humans were not only to be treated but also used as experimental subjects in the same way that mice currently were in the chemist's laboratory. And behind this idea, too, hovered the outline of a revolution in anatomy, with pneumatic researches generating new theories to explain how airs might diffuse from the lungs to permeate the body and expose the hidden mechanisms of life.

The idea of pneumatic medicine was, however, not entirely without precedent, and indeed Beddoes had done more than anyone over the previous few years to strengthen its foundations. When Joseph Priestley had, in 1774, produced the gas that would become known as oxygen, he had placed one of his laboratory mice in a bell jar filled with it, and had noticed immediately that it survived there for longer than in regular air. He was encouraged by this to breathe some of it himself, and recorded that he had felt 'particularly light and easy for some time afterwards'. 'Who can tell', he speculated, 'but that, in time, this pure air may become a fashionable article in luxury', just as soda water had?[34] Yet though Priestley's new gas had demonstrated plainly enough that the factitious airs could powerfully affect bodily functions, Beddoes was among only a very small minority of physicians who had attempted to put this discovery to practical use, or to pursue his experiments in the direction of a new chemistry of life.

He made his first breakthrough not in the laboratory but the library, in a long-forgotten 1668 tract by John Mayow entitled *De Respiratione*. Mayow, a Cornishman, had been a member of the Royal Society and a colleague of Robert Boyle, whose famous experiments on air had demonstrated that it was 'elastic' – capable of being compressed – and necessary for combustion. Mayow, however, had extended Boyle's work into physiology, discovering that a bellows inserted into a dog's lung would keep it alive as effectively as normal breathing, but that arterial blood from another dog was not sufficient to supply breath. The theory of respiration he developed was that life was supported not simply by air itself, as Boyle had assumed, but by particular corpuscles within the air which he called 'fiery nitre': these were the particles that permitted combustion, and also those which were extracted from the air by the lungs. Dusting off *De Respiratione* over a century later, after Antoine Lavoisier's definitive identification of the principle of oxygen, Beddoes was staggered by Mayow's prescience; he edited the tract for republication under the title *Chemical Experiments and Opinions Extracted from a Work Published in the Last*

Century, and hailed its author as a genius who deserved to stand by the side of Newton. Mayow, he claimed, 'aspired to change the whole face of medicine and physiology, by the application of his wonderful discoveries to the appearance of animal nature';[35] but while Newton and Boyle had succeeded in transforming physics and chemistry, the key to the transformation of medicine had been lost.

These insights highlighted the importance of respiration in ways that canonical eighteenth-century medicine had failed to grasp. The function of the lungs was still typically held to be an adjunct to that of the heart: keeping it pumping by their regular expansion and contraction, or cooling it by introducing air to the chest. If, however, respiration involved changing the balance of different gases in air, then inhalation, absorption and excretion of gases represented an invisible equivalent of food and drink, without which no life could survive. Over the previous decade, a handful of physicians had begun to sketch the contours of this potentially vast new territory. The great Italian experimenter Lazzaro Spallanzani, who had already demonstrated the existence of microbes and performed successful artificial insemination on a dog, had shown that oxygen was carried all round the body by the blood; in 1784 the twenty-five-year-old Beddoes had translated his tortuous dissertation with precocious elegance. In 1788 Beddoes' Edinburgh colleague Edmund Goodwyn's tract *The Connexion of Life with Respiration* had shown that respiration caused the blood to undergo chemical changes in the lungs, darkening its colour and, he speculated, imparting a force that powered the heart as the blood rushed through it. Such researches had led Beddoes 'to reflect with peculiar earnestness' on the mysterious dance between air, lungs and blood, and the therapeutic potential that might lie concealed within it.[36]

He was well aware, however, that to follow these researches was to swim against the medical tide. The world of medicine, particularly in Britain, remained a curious exclusion zone from the aspirations to scientific progress that had characterised the philosophies of the eighteenth century. The physical sciences were proliferating; anatomy and botany were exposing ever more delicate fibres and filaments, fluids and juices; the effects of stimuli and irritants on muscles and nerves were being ever more closely mapped. Yet the idea of applying chemistry to medicine struck most practitioners not as new and exciting, but as antiquated and crude. The cult of the chemical remedy was still associated with the 'iatrochemistry' practised by followers of the Renaissance alchemist Paracelsus, with its heroic, kill-or-cure doses of toxic minerals such as antimony and mercury. Such drugs remained the business of apothecaries and druggists, a species far more numerous than physicians, whose remedies were typically a cheap substitute for professional care. 'Blame', the great seventeenth-century doctor Thomas Sydenham had pronounced, 'lies at the door of those who have so tortured and overheated their brains to believe, that the chief weakness of medicine is its want of great and efficacious remedies'.[37] Most doctors used a small range of drugs – emetics, purgatives, cloves for toothache – that could be had from any small-town apothecary and that, as Sydenham dismissively observed, 'a druggist's shopboy can tell me offhand'.[38] The prevailing view was that only the quacks who preyed on the desperation of their patients exalted these props as panaceas; for Beddoes, however, the quack trade would only be marginalised once the treatments offered by those who had a proper scientific understanding of medicine could be distinguished from those proposed by people who did not.

There was a precedent for the modern synthesis of chemistry and medicine that Beddoes was attempting, and it was promising and problematic in equal measure. It was indelibly associated with the maverick doctor John Brown, an energetic and charismatic Scot from an impoverished rural background who had, like Beddoes, been drawn to study medicine at Edinburgh University. There, he had broken dramatically with his professors, and had started teaching his own theories alongside theirs. The Brunonian system, as it became known, rejected the entire edifice of learned eighteenth-century disease classification, diagnosis and treatment, and replaced it with a new and dramatically simple 'science of life', from which followed an equally simple regime of medicines that, he claimed, rendered professional medical training virtually redundant.

Brown's theory was rooted squarely in Newtonian mechanics and its all-embracing chain of material cause and effect. All human, animal and plant bodies, he argued, were permeated by a life force that burned like a candle in a process of slow combustion, supplying an energy that distinguished them from inanimate matter and that departed the physical body at the point of death. This force Brown termed 'excitability', and he located it in the nervous system, where it reacted with external stimuli to produce the complex of actions and reactions that we call life. The reservoir of excitability was depleted by physical and mental exertion, just as it was recharged by food and sleep. What we call health was excitability and stimulus in equilibrium; what we call disease was forms of imbalance in which the organism was either understimulated (with symptoms such as weak pulse, pallor or lassitude) or overstimulated (high blood pressure, fever or mania). In the Brunonian system, every disease could be understood as a point on this spectrum; Brown even dreamed of devising a thermometer that would reduce all ill-health to a single and measurable scale. By the same token, all diseases could be corrected by drugs that countered the imbalance: typically alcohol to boost excitability, and opium to reduce it.

Brunonianism pitched its camp in truculent opposition to the establishment, and in so doing took on the cast of a revolutionary movement. It generated popular tracts that spoke over the heads of the physicians, offering patients the means of taking charge of their own healthcare without costly intermediaries. It treated the medical profession as an emperor with no clothes, a self-serving elite and a secular priestcraft. It was a heretical movement, but one not without successes: in fastening on the two most potent chemicals in the doctor's bag and using them with unprecedented liberality, it was often more effective than the ministrations of its opponents. It appealed strongly to those who, like Beddoes, felt that orthodox medicine was marooned in the Dark Ages and who were seeking a modern, rational and scientific alternative. It had greater success in countries such as Germany, Austria and America, where it took root in the professions, was taught in universities and promoted by leading practitioners. To the mainstream of British physicians, however, it remained old-fashioned quackery in newfangled Newtonian garb which ignored the gentle efficacy of traditional Hippocratic therapies and boiled down to no more than advocating large quantities of brandy and opium for every ailment.

Beddoes had met Brown in Edinburgh in 1782, after which he 'never desired his conversation a second time':[39] having watched him take fifty drops of laudanum in a glass of whisky five times during the course of a lecture, he formed the opinion that his prescription for others was also the main form of sustenance for his own overexcited nervous system. Nevertheless, he believed that Brunonianism contained the seeds of a medical understanding that would continue to flourish and produce new chemical treatments long after the fashionable theories of the day had been forgotten. It was, in Beddoes' view, 'the only specimen of extensive reasoning in pathology, calculated to afford any satisfaction to a just thinker'. He stressed that 'I have always found full as much to reject as to receive' in Brown's views;[40] he held no brief for his personal conduct or the medication regimes of some of those who practised in his name; yet he would inevitably find himself bracketed with 'the Scottish Paracelsus', as much for his antipathy to the elitism of the medical profession as for his quest to transform medicine with new and powerful chemical agents. It was a characterisation that would help Beddoes' cause in some respects and hinder it in many others, but it was not an entirely unfair one. What Beddoes envisaged from pneumatic medicine was in many ways an extension of the Brunonian pharmacopoeia – and an elimination of its weakest point. Alcohol and opium were, after all, blunt tools with well-known limits: increasing their dose was widely understood to be a game of diminishing returns that eventually did more harm than good. Pneumatic chemistry, by contrast, promised a cornucopia of new stimulants and sedatives, delivered in concentrations unknown in nature, with therapeutic effects that could thus far only be guessed at.

*   *   *

Beddoes and Giddy stayed in Tredrea deep into autumn, until Oxford's Michaelmas term was well under way and Beddoes' teaching duties were upon him. On 30 October they parted at Camelford's turnpike gate, where Beddoes boarded a coach back to Devon; on 2 November he wrote to Giddy from Exeter, where he had succeeded in hunting down the latest *Gazette nationale*, news from France unadulterated by the 'established system of misrepresentation' of a British press who 'select everything that can bring discredit' on the Assembly's efforts 'and suppress the rest'.[41] He read that the revolt in Santo Domingo had been put down with two hundred casualties, an outcome 'more favourable than I had hoped'. He inhaled Devon's crisp autumn breath, 'her green meadows, her cultivated smells and her clear streams', a tonic to the senses after the wilderness of the far west.[42]

In an enclosure written in French, he also revealed to Giddy an additional and thus far secret source of the joy that had filled the summer: he had developed an attachment to his correspondent's younger sister, Philippa. He refers obliquely to a delicate business that he had not found an occasion to raise before, a fear of giving false impressions and of possible obstacles to the fruition of his hopes; clearly, Giddy's stiffness on romantic matters had made the subject impossible to broach face to face during his visit.[43] No previous attachments are recorded in Beddoes' life up to this point: although he had grown up comfortable in female company, his awkward physique and his intellectual appetites had conspired thus far to propel him through a bachelor's life in a succession of overwhelmingly male environments. But the summer in Cornwall had given an intimation that in this, as in much else, his life was on the point of change.

From Exeter, he travelled by coach up through Bath and back to Shropshire, with plenty to consider during the long journey. He was still mulling over his geological researches, and the use to which he might put them in his new series of lectures. He was aware, too, that he had presented young Giddy with some sensitive information that he might find hard to digest. But he was also considering an abrupt career change, and assessing the practicality of his scheme to act as midwife to the yet unborn discipline of pneumatic medicine. It was by no means a simple decision and, if it failed, would seem an inexplicably reckless one. He would be leaving a secure post at a prestigious university for a project without precedent. Chemical researches in general were almost exclusively the hobby of gentlemen with the means for a private laboratory: the only professional employment they offered was in the universities, where he was already well situated, or perhaps in industry, most likely grinding away at industrial processes in the clanking manufactories of Birmingham. And chemical researches within medicine, to the extent that they existed at all, were still widely associated with quacks driven by no ambition beyond finding plausible nostrums with which to exploit the poor and desperate.

Beddoes was not only well situated at Oxford, but in the process of transforming his subject in important ways. His chemical readership had, over the previous three years, become a striking and widely remarked success: he wrote to Joseph Black in 1788 that his last course had been 'the largest class that has ever been seen at Oxford, at least in the memory of man', and claimed elsewhere that he was drawing crowds on a scale that had not been seen in Oxford since the Middle Ages.[44] He was promulgating chemistry with an energy and effectiveness matched only by Black himself, and across a swathe of the nation's elite that Black, in Edinburgh, was unable to reach. Oxford's chancellor had expressed hopes that, with a man of Beddoes' calibre in the post, the king might approve the appointment of a regius chair in the subject, giving him a full professorship. He was on an exhilarating trajectory, one that would soon see him at the top of his profession, and with every chance of taking the profession itself to previously unscaled heights. In chemistry, as in politics, a French revolution was under way, and Beddoes was better positioned than anyone to sell it to the British scientific public.

The central figure in this chemical revolution was Antoine Lavoisier, and its conclusions had been advertised in 1789 in his *Traité élémentaire de chimie*; but, as with the political revolution, there were few in Britain who had thus far grasped its full significance. It was a revolution that encompassed sweeping changes in the experimental method, with new instruments capable of measuring with staggering precision; but its focus was the nature of chemical transformations, and particularly of combustion. Throughout the eighteenth century, this process had been conceived as the release of an invisible spirit of fire, named phlogiston, which was contained in varying quantities in different forms of matter. It was the release of phlogiston that caused burning matter to give off light and heat; when its fiery qualities exhausted themselves, substances shrank to their skeletal form of inert ash. Similarly, metals rusted as they released their phlogiston in a form of slow burning, and air saturated with phlogiston would no longer keep a candle alight.

Phlogiston had itself been a revolutionary doctrine when it was conceived, and had stood for a century as the first rational theory of chemistry's foundations. The term had been coined in 1697 by the German chemist Georg Stahl, for whom it liberated chemistry from the corpuscular and mechanical theories of Robert Boyle and opened up a new language for describing the mysterious ways in which pure substances, isolated and recombined into new configurations, could manifest properties that could never have been inferred from their original components. It was, in a sense, a return to older and less materialist forms of thought, a refinement of the Paracelsian idea of alchemical sulphur, the principle of fire, which was not a form of matter but rather a spirit, or temperament, with which matter could be imbued. Yet for all its imponderable qualities, phlogiston had proved a useful and timely innovation, and had enabled chemistry to make important material advances. It had provided a shared language within which the previously scattered and unsystematic disciplines of pharmacology, mineralogy, alchemy and metallurgy had coalesced into a coherent whole; and it had put an end to the profusion of elaborate suppositions about the properties of invisible particles, focusing the new science on measurable forces.

By the end of the century, however, phlogiston was becoming a victim of its own success. The march of experimentation, observation and measurement had inevitably exposed its properties to question: when metals were burned, for example, they produced an ash, or calx, that weighed more than before. To make sense of this, some had suggested that the escaping phlogiston was lighter than air, others that it had negative weight. Such assumptions were problematic but, for most, tolerating their contradictions seemed preferable to tearing down the scaffold of theory on which the chemical community was working. Other phenomena – light and electricity, for example – were agreed to have material form but no measurable weight. It was Lavoisier who took the decisive step, announcing that phlogiston was no more than a ghost in the machine. In 1772 he had begun slowly burning sulphur and phosphorus and carefully measuring their increase in weight, which had convinced him that they were not shedding phlogiston but taking up some invisible component from the atmosphere. In October 1774 he had had a visit from Joseph Priestley, who told him excitedly about the new gas he had just isolated in which candles burned more brightly and mice lived longer. Lavoisier had conducted his own experiments with the new gas and, by March 1775, had announced his discovery of oxygen, demonstrated that it was the substance consumed in combustion, and lodged his claim of priority in a sealed envelope deposited at the French Academy of Sciences.

Where Priestley saw himself as part of a communal endeavour to elucidate the mind of God, for which no man deserved credit, Lavoisier was predatory in claiming such discoveries for his own: the Swedish chemist Carl Scheele had an equally good claim to having discovered oxygen, but his researches too were annexed by Paris. Lavoisier was focused on theories, and on the sequence of experiments necessary to demonstrate them, in a manner that none of his competitors could match. Priestley was a prodigious experimenter and discoverer, but never entirely coherent in systematising his results: his designation for oxygen had been 'dephlogisticated air', on the grounds that the weight of his mercury calx must have drawn the phlogiston from the air in which it burned. As Lavoisier raced ahead with the implications of his discovery, formulating a table of elements to which the many compounds that he had now shown to be 'oxides' could be further reduced, Priestley dug in his heels, and insisted that the new theory created as many problems as it solved. Many British chemists joined Priestley's new dissenting denomination of 'Phlogistonians', and chemistry began to bifurcate, with discoveries on either side of the Channel wrapped in language that made each incompatible with the other.

Beddoes, unlike most of his colleagues, had been intimately involved in this debate for several years and was thoroughly conversant with the new language. He had, as in much else, followed in Priestley's footsteps by visiting France in the summer of 1787, where he had met Lavoisier and studied in Dijon with his collaborator Guyton de Morveau, who had been impressed and delighted by the young English philosopher; they had worked together in Morveau's laboratory, and celebrated the Burgundian festival of the new vintage together in Dijon's medieval streets. Morveau had just begun working with Lavoisier on the new table of elements, and on developing for it a universal nomenclature within which, as in Carl Linnaeus' taxonomy of the vegetable kingdom, all new discoveries could be contained. Unlike Priestley, Beddoes' travels had converted him to Lavoisier's system, and to the necessity of British science adopting the language in which the next century's chemistry would plainly be conducted. It was a task for which his new Oxford lectureship positioned him impeccably.

It was, however, no simple prospect. For those who had grasped Lavoisier's elegant logic, the chemistry that had accreted over the previous century now seemed a hopeless tangle. It was not merely the language of phlogiston that needed to be purified: the nature of Lavoisier's revolution meant that there could be no simple translation from the old to the new. Some British chemical names built on residual terms from alchemy, such as vitriol, spirit, butter, liquor or flower for different types of essence and compound, or the abstract principles of salt, mercury and sulphur. Others, such as Joseph Black's 'fixed air', referred to particular properties that they had manifested in the experiment that had first isolated them. Pneumatic chemistry, which had grown in piecemeal fashion over the previous fifty years, was a cloud of confusion: the word 'gas', coined by the seventeenth-century Paracelsian Johann Baptista van Helmont from the Greek term 'chaos', coexisted alongside the generic term 'air' – or 'factitious air', for a gas not found in nature – as well as 'elastic fluid', a designation that had its roots in the corpuscular theories and air-pump experiments of Robert Boyle. Not only French, but German and Italian had their own linguistic conventions; even the nomenclatures taught in London and Edinburgh were increasingly divergent. To render this babel intelligible would require a master of chemistry and its many tongues.

Yet there was, in Beddoes' view, more to be done than simply translating and codifying Lavoisier's theories. France's chemical revolution, like its political one, was still unfinished, the ferment from its wild gas not yet settled. The new theory had not, on closer inspection, exorcised the ghost in the machine as cleanly as first appeared. Lavoisier had initially christened oxygen a 'principle', and had retained in his table of elements a place for light and for 'caloric', a hypothetical fluid present in air: combustion, in his new scheme, was the process that allowed light and caloric to be released as flame. His naming of oxygen from the Greek word for 'acid-maker' expressed his conclusion that it was a constituent of all acids: a reasonable induction, but one that would prove to be false. 'Till the sciences shall be considered perfect', Beddoes wrote to Joseph Black on his return from France in 1787, 'successive changes of method will be requisite'.[45] He sensed that there was a further revolution to come. He was correct, and indeed he would be instrumental in bringing it about.

* * *

Such was the prospect that Beddoes had before him on his long journey back to Shropshire, and against which his new scheme would need to be weighed. His current career held many advantages, and might be turned to still greater ones. As well as championing the new chemistry, he was also planning to spin his summer researches into a new lecture series on geology, exposing Oxford's students to the abysses of deep time that had thus far been glimpsed by so few. But these possibilities were balanced by disadvantages that had, over the last year, come to bear more heavily on him. The professorship now being dangled before him was a salaried position, but his current post of chemical reader, though prestigious, was unpaid: incumbents were expected to make their living from charging attendance fees to their lectures and from extracurricular work, such as private tuition, publishing, medical consultation or stipends from other formal posts held simultaneously. The laboratory facilities at his disposal were also poor: a dingy basement with basic furnaces that very few of the crowds to whom he lectured would ever visit. Beddoes had supplemented them with some modern equipment – including troughs and vessels, like Priestley's, made to order by the Wedgwood factory – but his experimental facilities remained less well equipped than many private laboratories to which he had equally convenient access.

In teaching modern experimental chemistry, too, Beddoes was coming to feel that he was swimming against Oxford's cultural tide. Prior to the reforms of 1800, the university was still essentially an Anglican seminary, attendance at which was a qualification for holy orders. All students had to profess the Thirty-Nine Articles of the Anglican church, and attendance at divinity lectures was a compulsory element of ordination; Beddoes' lectures, by contrast, were essentially recreational, and serious study in chemistry would involve further years at a specialist university such as Edinburgh. He had already published a complaint about the Bodleian Library, whose slackness in regard to opening hours, ignorance of new books and obstructive staff made it, in his view, inadequate for the serious acquisition of modern scientific knowledge.[46] He might be introducing chemistry to some of the best and brightest of the new generation, but he would always be operating in a climate where experiment was distrusted and facilities were second rate and where he would pass his most promising students on to others to complete their education.

The events of the summer had also brought home to him that Oxford's tenaciously held traditions were making it a bastion of reaction in a fast-changing world. The new German criticism of the Bible, for example, was strictly policed: historical studies of Judaism in the Roman Empire, or collateral evidence relating to the separate tradition of the Old Testament psalms, were still regarded as inadmissible. Non-Anglican Christianity was a source of suspicion, and when combined with political reform it was seen as positively dangerous: George Horne, the president of Magdalen College, regarded the likes of Joseph Priestley as enemies of religion, the modern equivalent of 'the Sadducees of Jerusalem'.[47] The Priestley riots in Birmingham would, Beddoes knew, have hardened these opinions, and he may have delayed his return partly on this account: if he suspected that his sympathy with Priestley's combustible mixture of chemistry and politics would cause friction in the year ahead, he would be proved right. His tentative attachment to Philippa Giddy also signalled, perhaps, a desire for a new and more expansive life beyond Oxford's largely celibate society. Although his prospects there remained bright, there were also many reasons to conclude that it was time to move on.

*   *   *

Davies Giddy returned home from Camelford to a stack of letters from Beddoes, who was still buoyed up by the exhilarations of the summer. 'I do not believe,' Beddoes confessed, 'that I have ever written so much about myself as I have lately done to you'. En route back to Shifnal he had stopped over in Birmingham, where the events of Bastille Day were still simmering in a state that seemed to him not far short of civil war: 'you can hardly form an adequate conception of the fanaticism of the people'. He had stopped to converse with 'several friends of Church and King and what was no trivial penance I hearkened to their political sentiments'. Their ideas, he informed Giddy, 'resemble exactly a mass of felt. It would be an entire loss of labour to disentangle it and put them straight, and if you once had got them so, they will immediately return by their own natural rigidity to their former confusion.'[48]

Giddy was in the midst of preparing for his return to Oxford, where he expected to catch up with Beddoes in a few weeks; but he was interrupted by an entirely unexpected development. Just as he was on the point of leaving, he found himself nominated sheriff of Cornwall.

This was an appointment of the Crown, made from a list submitted by the county's judges: in practice, the name at its head was always chosen. It made Giddy officially 'the first man in the county', technically superior in rank to the nobility; it also made him responsible for policing and public order. It was a great honour, and it was unthinkable that he would disappoint his family by refusing it; but it was also a shock, and not an entirely agreeable one. He had no idea why he had been offered it, but he could immediately see that it promised to be a poison chalice. It would be his specific brief to prevent riots: in the type of situation that had boiled over in Birmingham, it would be his duty to give orders to the local and county militias, and it would be he who would be responsible if the situation escalated into violence. In such a crisis there would be no room for ambiguity in his political views.

It would be thirty years before he got to the bottom of why the appointment had fallen to him; even then, the picture he assembled would have its contradictory elements, but it would be clear enough that his summer with Beddoes had not passed without comment among the Cornish establishment. The prime mover in the appointment was most likely Sir Francis Bassett, who had been bitterly angry at Beddoes and Giddy's criticism of his plan for a county infirmary. He had hoped that the presence of a distinguished Oxford chemist would give a useful imprimatur to the project; instead, it had been shot down with a brusque polemic on the medical profession's systemic failure to treat the poor. The message that Giddy's appointment as sheriff was designed to send to a young Cornishman at Oxford was that with a respectable education and influence over public affairs should come responsibility, and not the luxury of playing revolutionary politics from the sidelines. If he wished to choose Priestley's politics over Burke's, let him take the consequences squarely on his own shoulders.

But there was also in the appointment a generosity and discretion that Giddy would only come to appreciate at a much later date. Flaunting his tricolour cockade had made a more pronounced impression on the county than he had realised at the time. It was becoming obvious to many that the faultlines running through British politics were likely to become far deeper, and that glib political gestures might be recalled later, with serious consequences for their authors. In Beddoes, Giddy had chosen a teacher who, whatever his other virtues, had little interest in his own political reputation, and was not to be trusted with that of his student. 'I had taken up the cause of France', Giddy recognised in 1826, 'with the enthusiasm which so just a cause is expected to produce on a mind that had ever been responsive to the call of Liberty. I might therefore have been led into joining some resistors and perhaps ruined.'[49]

But such insights were for the distant future. Giddy was strongly aware that he was being coerced into a role, and perhaps a life, not of his own choosing. As a consequence, his Oxford days were over: he would for the foreseeable future be confined to the remote province of his birth, just when he had been on the point of spreading his wings. At his investiture the following February he would offer at least symbolic resistance to his destiny. The sheriff was obliged to take an oath to uphold, among other things, the Test Act, which Giddy described privately as 'a disgrace to the nation';[50] at the same time he made a point of passing a copy of Tom Paine's *Rights of Man* to one of the magistrates, in full view of the assembled company. This was a gesture that within two years would be declared an act of sedition, punishable with long prison or transportation sentences, and Giddy realised almost immediately that 'the circumstances brought me into much discredit'.[51] But he remained stubborn, if not in his espousal of the cause of revolution, at least in keeping faith with the ambivalence he had felt while watching Burke and Fox debate the previous summer. He was not yet willing to become a partisan of either side.

Events would force him to choose soon enough. Like Beddoes, he was facing a long and testing struggle to remain true to his causes and to himself as he negotiated the tumultuous decade ahead. The friendship the two men had forged over the summer would prove as strong as any that either would ever make, and it would survive everything that the future would throw at them; but never again would they be able to take the same side in public under the bright colours of the revolutionary cockade. The Priestley riots marked a watershed, and their summer had been a brief state of grace before its consequences consumed them. They had worn freedom's garland for a season, but it was a season never to return.

2

The Lunar Son

E n route home to Shropshire, Beddoes took the opportunity to pay a visit to James Keir, one of the few men in Britain who could claim to be a professional experimental chemist. Keir had studied chemistry at Edinburgh twenty years before Beddoes and, after a period of military service that had taken him to the West Indies, had settled in the Midlands to turn his studies to commercial use. He had begun experimenting with the production of strong alkalis, caustic soda and potash, which were in high demand for manufacturing, glass-making, bleaching and fertilising; he had subsequently set up a chemical works that had pioneered new processes for soap-manufacturing. His advice on chemical equipment had long been invaluable to Boulton and Watt at their Soho works, and he had moved to Winson Green on the edge of Birmingham to supervise their glass manufacture.

Confident, ingenious, sociable and restlessly practical, Keir was a long-standing member of the Lunar Society who shared Beddoes' ambitions not only for chemistry but for politics, in which he was a firm supporter of Joseph Priestley's ideals of progress. Though far less comfortable than Priestley with the political limelight, he had found it turned on him over the summer since he had chaired the Revolution Society dinner in Birmingham that had sparked the Bastille Day riots. Now he was caught up in the inquiries that had been commissioned in their aftermath, which were turning up evidence that the Birmingham magistrates had colluded in the early stages of the violence, and even claims that the mob had been been deliberately incited by government *agents provocateurs*.

He and Beddoes were each eager for the other's news: Beddoes on the emerging political scandals that were receiving no coverage in the national press, and Keir on Cornwall's granites and basalts. The latter had a keen interest in geology: he had done numerous experiments on the local rocks of the Midlands, and shared Beddoes' conviction that the crystals in granites had been formed in molten lava as it cooled. He also prevailed on Beddoes to help him with the signature project to which he had devoted much of his life: a dictionary of chemistry that he had first published twenty years previously and that he was now overhauling for a new edition, the first part of which had already been published. Beddoes had already brought several entries up to date, but their collaboration had been made delicate by the phlogiston theory to which Keir, like Priestley, still adhered. They discussed phlogiston and oxygen at length, but could only agree to disagree: Keir accepted that Lavoisier's phosphorus experiments had proved that oxidisation took some ingredient from the air, but was unwilling to follow this logic to a conclusion that would be so destructive to his life's work. Beddoes was painfully aware that Keir's *magnum opus*, whatever its other virtues, would be hopelessly out of date before it was finished, and that Britain would still be awaiting the grand work of synthesis that could stand as the bible of the new chemistry.

Beddoes arrived home at the end of November to find his sister Rosamund taking tea with James Sadler, the balloon pioneer with whom he had collaborated on his simulation of a flaming meteor and who had subsequently become the technical operator at his Oxford laboratory. Over the previous three years, Beddoes had drawn Sadler into his chemical researches and had engaged him to upgrade the university's equipment: he considered him 'a pastry cook of this place, a perfect prodigy in mechanics' (Sadler was also a confectioner).[1] Now, Beddoes was helping him in return. For some time the mechanical prodigy had been tinkering with an engine that he believed would be superior to Watt's, and Beddoes and Giddy had been acting as intermediaries between his experiments in Oxford and the smelting and metal works of the ironmaster William Reynolds. Before setting off for Cornwall, Beddoes had brokered Sadler's passage to Shropshire to begin building his new engine prototypes at Coalbrookdale.

William Reynolds was a contemporary of Beddoes who had also studied under Joseph Black and immersed himself in Priestley's experimental chemistry; he was a direct descendant of Abraham Darby, the founder of Coalbrookdale's ironworks, and a scion of the wealthy Quaker dynasties with whom three generations of the Darby family had subsequently intermarried. His father, Richard, was now ironmaster of Coalbrookdale itself, while William and his brother

Joseph ran the works at nearby Ketley, which they had recently improved with a system of plateways, a pioneering design of iron tracks on which the ore wagons now ran. It was with Reynolds that Beddoes and Giddy had observed the transition from cast to malleable iron, and it was Reynolds who had pressed Beddoes to work his findings up into papers for the Royal Society. Though he was now a busy industrialist, Reynolds' interests still ranged widely, and he paid close attention to Beddoes' chemical and geological researches. For him, such experiments and theories were inseparable from the practical questions faced by mining and the new industries, and from the broader transitions in society that they were in the process of engineering. 'Speculations respecting the theory of the earth', as he put it, 'may appear to some of little consequence, but they are connected with and lead to speculations which are relevant to the arts & manufactures upon which depend many of the comforts of the more polished parts of this globe'.[2]

The society of Keir and Reynolds was congenial to Beddoes in ways that that of his Oxford colleagues was not, and it would be to their densely interwoven circle of friends and associates that he would gravitate over the coming year. Although he would return to Oxford to deliver his lectures, and continue to work his new ideas on chemistry and geology into academic form, he would spend most of 1792 in Shropshire, and by the end of it he would be ready to reinvent himself as a practitioner of applied chemistry in a field hitherto unexplored. His Shropshire was a world away from Oxford not merely in its respect for science, but in its broader attitudes to society and politics. William Reynolds, for example, although his family had been the wealthiest in the area for generations, remained conspicuously outside the local hierarchies of church, aristocracy and parliamentary influence to which, in most other parts of Britain, he would have been obliged to pay allegiance: he doffed his broad-brimmed Quaker hat to no one, and refused on principle to pay Church of England rates. In its quietist way, the dissenting world within which he had grown up was a state within a state, with its own schools, hospitals and asylums as well as its own churches and businesses, and it had been engaged for as long as anyone could remember in its attritional struggle to acquire the same civil rights as the rest of the population. It was a world that would offer Beddoes sanctuary from the forces of reaction that would sweep the rest of the country through 1792, though perhaps not to the degree that he assumed.

* * *

Although Beddoes' family were also in Shropshire, his relations with them had for some years been under strain and he would not remain long under their roof. In contrast to his childhood recollections of his adored grandfather who had made the family fortune, his relationship with his father was fraught and uncomfortable. Thomas senior had left his son, Richard, to manage the lucrative tanning business that he had built up, but he had strongly believed that his grandson should be educated to the full and left free to take whatever path his talents should open up to him. Richard had maintained the family business, but without his father's gift for enterprise: he had not wasted his fortune but neither had he expanded it, or stepped beyond his father's shadow. He himself had never had any ambitions for young Thomas beyond educating him at home and training him up to succeed him in the tanning business, and it had become increasingly galling to him to fund the succession of university degrees, doctorates and unpaid lectureships whose effect had been to maintain his son as a perpetual student far into adulthood.

As Beddoes grew to maturity, wrangles over allowances developed into bigger arguments about the sale of land and the transfer of a final inheritance. In letters to his mother, with whom his relations were more open and affectionate, Beddoes attempted to assuage 'all anxiety about taking too much of my father's property', insisting that 'it cannot be better laid out than in establishing me'.[3] But these disagreements were too raw to make an extended stay in the family home an agreeable prospect. William Reynolds invited him to come to his residence, Bank House in nearby Ketley, a plain but extensive mansion equipped not only with a fine library and fossil collection but with a chemical laboratory as well equipped as Oxford's. It was from this new base that Beddoes began to consider the logistics of establishing himself as an independent operator at the crossroads of chemistry, medicine and industrial enterprise.

Perhaps the closest and certainly the most successful model for the career he had in mind was provided by Erasmus Darwin, a pivotal figure in the Lunar Society and in many ways the presiding genius of the circle in which Beddoes was now moving. Darwin was, like Beddoes, a prodigy and polymath, but with a thirty-year start on his protégé. The scientific subjects over which his mind now ranged formed a vast tapestry that allowed him to conceive a natural history of the universe more intricately imagined than any before him. The cosmos had, according to Darwin, begun with a Big Bang that had 'exploded at the same time and dispersed through infinite space', and his investigations into the physics of whirling gases had revealed to him how these might have condensed to form stars and planets. He had studied the action of currents and condensation in clouds, wind and water, and the geology of coal, which convinced him that it had been formed over 'millions of ages' from compressed plant material; mountains were for him 'mighty monuments of past delight', a testament to the reproductive exuberance of the millions of generations of bivalves and crustaceans of whose shells they were formed.[4] He had been the first to elucidate the full chemistry of photosynthesis, and had traced the emergence of animal life through the successive stages of fish, amphibians and reptiles to humanity itself, which he saw as no more and no less than an animal form that had emerged over aeons through the shaping pressures of existence.

In many of these fields Darwin's expertise was eminently practical. He had been a pioneer of pneumatic chemistry, establishing in 1763 that the density of gases is a function of their pressure divided by their temperature, a discovery put to use by his friends James Watt and Matthew Boulton in their steam-carriage designs. His studies of botany had led him to recognise the value of artificial plant nutrients, including nitrate fertilisers. He had a knack, almost a compulsion, for ingenious inventions, and had devised electrical machines, steam turbines, wind-gauges, artesian wells and drills, telescopes and submarines, often doodled after a few moments of apparently idle daydreaming: his sketches included futuristic wonders recognisable today as internal combustion engines and even hydrogen-powered cars. His inventions were a manifestation of his confidence that the technologies with which he was tinkering were the harbingers of a happier world to come, and, as James Keir would later observe, 'the communication of happiness and the relief of misery were by him held as the only standard of moral merit'.[5] This conviction made him a freethinker who distrusted dogma and hated oppression in all its forms: he supported his friend Benjamin Franklin in championing American independence and opposing slavery, and found it equally natural to approve of female emancipation and the French Revolution. Like Beddoes, his expansive mind was complemented by a nearly spherical form, and he had famously had a semicircle cut out of his dining table to accommodate his stomach.

Like Beddoes, too, the discipline that underlay all of Darwin's learning was medicine. By 1791 he was perhaps the most famous doctor in the country. He had turned down requests from George III to become his personal physician, and his polymathic interests were pursued between long carriage rides on rutted country roads, paying home visits to the patients who clamoured for his services from every corner of the country: by his own calculations, his practice involved travelling over ten thousand miles a year. The wealthy paid handsomely, while the poorest received their treatment free. He had also taken his medical degree at Edinburgh, and his friendship with many of Beddoes' new circle, including James Keir, dated from those student days. Beddoes had first met him, probably through Reynolds, at Shifnal in 1790, and they had been corresponding ever since, mostly about geology. Beddoes had written to him earlier in 1791, proposing 'an exchange of the Shropshire for the Derbyshire fossils' and asking how long ago he believed the coal and tar in the Coalbrookdale pits was formed and to what depth it might extend.[6]

On his second marriage Darwin had moved north from the Midlands and closer to Shropshire, and the correspondence and visits between the two now became more frequent. Of the many things that Beddoes wanted to learn from Darwin, the most pressing was how to build up a medical practice lucrative enough to support the type of scientific enterprise he had in mind. He had already begun to expand his services as a doctor, and he would spend a great deal of 1792 bumping up and down in carriages across Shropshire, North Wales and beyond. Darwin, who was at this point earning over £1,000 a year as a physician, advised him on fee scales and reminded him that travel would occupy at least as much of his time as doctoring and, especially with wealthy patients, should be charged accordingly.

Beddoes was inspired, too, in another capacity. Darwin had always possessed a fluency in Augustan verse couplets, and had often worked visions and amusing skits into elegant classical form. For most of his life these had been no more than *jeux d'esprit* for private consumption, but in 1789, at the age of fifty-seven, he had published a verse epic entitled *The Loves of the Plants*, a work in four cantos totalling nearly two thousand lines, interspersed with dialogues and buttressed with footnotes that, in elegant form and genial style, introduced the reader to the entire plant kingdom, arranged according to the underlying reproductive structures that formed the basis of the Linnaean taxonomy. The work appeared anonymously – Darwin had been concerned that it might appear too frivolous and racy for a doctor – but with its overwhelmingly enthusiastic reception his authorship became an open secret, and he announced that it constituted only part of a continuing project entitled *The Botanic Garden*, of which further volumes were to be expected.

But *The Loves of the Plants* alone was enough to make Darwin the most popular British poet of the day. It embodied the most up-to-date understanding of the natural world that had ever been offered to the general public, presented not as a textbook or treatise but as an intimate view of the drama of life exquisitely rendered in verse. Through the vast sequence of nature epics that *The Loves of the Plants* began, Darwin would build on classical ideas of life arising in primitive forms and metamorphosing into the dazzling variety we see today: of blossoms exploding from buds, caterpillars transforming into butterflies and tadpoles into frogs; of the primitive dog bred for speed into the greyhound, and strength into the bulldog; of strange forms and mutations being passed on to offspring, and curious features such as the elephant's trunk and the parrot's beak emerging as a response to their peculiar environmental challenges. Although his notion of the mechanism by which such changes came about was closer to the purposive struggle later proposed by Jean-Baptiste Lamarck, the vision of life's grandeur that Darwin unfurled was one that would become for ever associated with his yet-unborn grandson.

Beddoes was delighted by the playful and inclusive manner in which Darwin's poetry insinuated these images, along with a prodigious quantity of hard science, into the public arena. Implicit within the logic that they unfolded, as within the geological story that Beddoes was assembling, was a timespan vastly greater than most people had ever attempted to imagine; but the act of doing so had, in Darwin's hands, become not a terror but a wonder and a delight. Beddoes himself had always had a knack for turning out verse, and was soon working on his own 'botanical dialogues': now, he confessed to Giddy, he 'would give anything I have to give, to show them to a certain person, and to hear the opinion of that person'. Beddoes was limbering up for a year of great productivity as a writer, and the pleasure of versifying seems for him to have been the facility with which he found the lines flowing. 'I write them', he told Giddy, 'as fast as I write this letter'.[7]

Alongside doctoring and poetry, Beddoes was also working with William Reynolds on Sadler's engines, two of which were now running at Coalbrookdale, and on Reynolds' attempts to strengthen his bar iron by resmelting it with manganese, a process mastered by the closely guarded Swedish Bessemer process whose standards Reynolds would approach by the end of the decade. He was also becoming active in local politics. The Society for Effecting the Abolition of the Slave Trade in Shropshire was a busy organisation that included Darwin and Reynolds among its members, and in early 1792 it was working overtime to muster support for the abolition bill that was shortly to be brought before the House of Commons. At the first meeting that Beddoes attended, he read out a petition that he had drafted 'by way, he said, of a confession of his faith', and asked for a copy of the final draft on a large piece of parchment to take to Shifnal for signatures. The organisers were sceptical about extending the petition to small towns, but Beddoes argued that it was essential to show 'the universal voice of the nation', and he returned the parchment with 150 signatures. The combined petitions, by the time they reached Parliament, stretched to 150 feet in length.[8] The Society published the evidence presented to the House of Commons, and Beddoes encouraged an ethical scheme to replace slave-produced West Indian sugar with Chinese imports and American maple syrup, and thereby 'undermine an evil which our legislature has not virtue enough to extirpate'.[9]

It was becoming a very full, even chaotic, portfolio of causes and activities, especially for someone whose primary employment was still as an Oxford lecturer. Beddoes' energies and enthusiasms, always compulsive, were propelling him in every direction at once. If he was to abandon the academies in which he had spent his entire life and enter a commercial world of which he had no experience whatsoever, he was well aware that he had much to learn, and fast. But the vast number of fields that he was attempting to master simultaneously seems at times to have generated a mild form of mania. In addition to writing and doctoring, experimenting and campaigning, his letters to Giddy detail ever more schemes on the boil, and spectacular new discoveries glimpsed around every corner. He moves within a single sentence from 'I am in some hopes that I have fallen on a plan of curing the diabetes' to 'I think I have fallen upon a scheme which, if it would be executed, would make perfect iron'.[10] The impulsive quality of his thought is further reinforced by an abrupt retraction of the sentiments he expressed towards Giddy's sister Philippa after his departure from Cornwall. 'I wrote to you some time since in bad French', he announced in March of 1792. 'There was nothing of the smallest consequence in the letter. I suppose I wished to hide its emptiness under the veil of a foreign language. How often has Latin served this purpose?'[11] Topics and schemes course through his letters with a profound restlessness: he is also contemplating a visit to Paris, to obtain at first hand the impartial news that has now become impossible to come by in Britain. Among all his other preoccupations there is mention, too, of investigations, carried out with Reynolds, Darwin and others, into the logistics of researching and practising pneumatic medicine. But before any such scheme could be considered in detail, it would be necessary to resolve the question of his future at Oxford.

*   *   *

Beddoes was not short of material for his Oxford students. His researches in Cornwall had provided a stock of new findings for his geology lectures, and his students were beginning to provide more. One had made his own visit to the borders of Devon and Cornwall, and had located the kind of site for which Beddoes had been searching: not merely a stratum of igneous rock but 'the aperture whence the glassy lava issued', a volcanic plug showing where granite had erupted through slate bedrock, and thus 'clearly indicating the posterior and igneous origin of this supposed primitive material'.[12] Beddoes was now reading a course of lectures on the natural history of fossils, which he hoped to work up to fill the critical and conspicuous gap in the book market for 'a popular and not altogether a superficial view of Hutton's subterranean system'.[13] Further notes show a well-polished series of lectures on the nature of strata and their implications for the forces that have acted on the earth over time, initial notes on atmospheric electricity and Benjamin Frankin's experiments with lightning, and on the vegetable origins of coal and the pressures and temperatures required for its formation.[14]

Beddoes was, however, visibly running out of enthusiasm for the university itself, and was spending little more time there than the minimum required to attend his own classes. He had already moved his technical assistant up to Shropshire, and Reynolds' laboratory had replaced Oxford as his experimental base. He was still proudly relaying tales of his overflowing lecture halls to Darwin and others, but privately he was beginning to feel that chemistry at Oxford had passed its peak. He had arrived in a scientific backwater, and had succeeded in demonstrating that chemistry was 'neither a petty branch of medicine nor one of the black arts'; but now that this message had been received, 'the stock of curiosity seems nearly exhausted',[15] and he was coming to believe that his studies were doomed to shrivel 'under the shadow of ecclesiastical & scholastic institutions'.[16] Society in Oxford was oppressive: wearing cockades and commemorating Bastille Day had both been officially banned, and the university's chancellor, Lord North, proclaimed himself at every opportunity a zealous defender of Church and King and a sworn enemy of revolution.

In private Beddoes was disillusioned by the progress of the French Revolution. He was dismayed by the factional infighting that was consuming the French Assembly, the Feuillant and Jacobin extremes holding the moderates to ransom with thinly veiled threats of mob violence, and the people's rejection of their early unifying hero, Lafayette, now that troops had been sent out against the crowds on the Champ de Mars. In his letters to Giddy he excused the Assembly by enumerating the obstacles to peace and progress that it faced: the chaos, the lack of information, the rise of fanaticism and faction, the hostility of foreign monarchies. In public he repeated the fragments of positive news that he could glean, mostly from the foreign press: how quickly and efficiently, for example, the lawsuits over property were being settled in the French courts ('more than 4000 in 11 months in Paris alone. More than one half of the whole number!').[17] Unless and until the revolution entirely betrayed its original intent, he would feel duty-bound to support it against its enemies. To do otherwise would be to reject the very idea of progress by experiment, and to surrender to the dead hand of authority and tradition.

But defending the revolution was becoming more than an abstract position of principle: it meant defending the Jacobinism with which it was becoming synonymous, at least in Britain. The second part of Tom Paine's *Rights of Man*, published in February 1792, had extended the logic of support for the French experiment into an explicit call for the abolition of all monarchies. The Jacobins were now equally explicit in calling for an international war against tyrants everywhere, and for the infiltration of revolutionary fifth columnists into the British masses. Cheap sixpenny editions of Paine's work were selling out as fast as they rolled off the presses, reaching a readership far larger than had, up to this point, been considered the entire book-buying public. These readers, most from a class of society thus far assumed to be beneath political debate, were meeting in London and provincial towns across the nation to read the book and to call for political reform, their numbers proliferating through a cell structure that sprouted new organisations like Hydra's heads. In May the London Revolution Society published its correspondence with the French Assembly and the Jacobin clubs, which had been the original spark for Burke's *Reflections*; it hoped by doing so to demonstrate the innocuous nature of its activities but, in the new climate, the revelations seemed to give chapter and verse to Burke's worst fears that the shadow of Jacobinism was stretching over Britain. The clamour for government action grew and, on 21 May, the home secretary, Henry Dundas, issued a royal proclamation that declared all revolutionary writings subversive, and announced that Tom Paine would be prosecuted for seditious libel.

Against this background Beddoes' politics were making him conspicuous in Oxford, and drawing him into battles that he had no wish to fight. Already the Home Office was beginning to assemble lists of political undesirables, among whom university teachers were prevalent. It was not as if Beddoes had no alternative: he had another home and another career, and no shortage of other projects jostling for his attention. Whether politics was the cause, or a pretext, or simply one reason among many, in June he offered his resignation to Oxford's vice-chancellor, John Cooke.

But it was not accepted: instead, he received a counter-offer that revealed delicate machinations in progress of which he had been unaware. The university was in discussion about applying for a regius chair in chemistry, and Lord North was on the point of putting the case to the home secretary. Cambridge already had a chair in chemistry, which was supported with a grant of £100 a year from the Exchequer; and chemistry, for all Beddoes' doubts, was coming to be recognised as an important and fast-developing field of knowledge, for which laboratory equipment and supplies were crucial and which could not survive on a shoestring budget for ever. One of the strengths of Oxford's application for funding, it was felt, was that the current reader was already acknowledged as a leading figure in the discipline. Cooke put it to Beddoes that this was the worst possible time for him to leave, and that many of his long-standing concerns were on the point of being addressed. As a regius professor he would have not merely the title he deserved, but a salary, library and laboratory to match.

Beddoes responded with a compromise: an offer of delay. Even if he had made up his mind to leave, it would still take time to pack up his materials, mineral collection and laboratory equipment, and he had no objection to delivering another set of lectures while he considered his future. He agreed at least to delay his resignation until the following year, and promised that, if the numbers in his class did not begin to decline as a result of his political profile, he would continue as chemical reader until the following summer. On this basis Cooke supported the recommendation that the chair, if it was granted, should be awarded to Beddoes, who had 'so unequivocal a claim in his line' and had 'given such ample satisfaction by courses of chemical lectures to a larger class of pupils than ever was before collected in that branch of useful science' that there was neither point nor justice in looking for any other applicant.[18]

This was how matters stood when Beddoes returned to Shropshire for the long summer break of 1792. It was a neatly engineered compromise that suited all parties and kept all doors open, but by the time he went back to Oxford it would be irrelevant.

* * *

The royal proclamation against seditious writings had direct consequences for Davies Giddy. It required assent from every county, and he, as sheriff, was obliged to convene a meeting of magistrates in the county town of Truro to read it and record a vote of loyalty to the king. It was not lost on him that, in practice, the target of this ritual was not the alleged sedition-mongers but those in authority who might wish to take a more liberal line on political dissent, or that many of his colleagues were well aware of the uncomfortable position in which it placed a sheriff who had brandished Paine's book at his investiture. 'Notwithstanding that I totally disapprove of the late proclamation', as he confided to his diary,[19] he was forced to intone a solemn warning against seditious writings 'circulated for the purpose of exciting groundless discontents and jealousies in the minds of your Majesty's loyal subjects', and to 'beg leave to assure you of our detestation of such proceedings, and our abhorrence of such correspondences'.[20]

There were several at the Truro meeting who were determined to play out this piece of theatre for all it was worth. One of the Cornish Members of Parliament, Francis Gregor, made a point of reading lengthy extracts from Paine and other works of which he knew Giddy approved, pausing frequently to denounce their sentiments. When he read a passage expressing Paine's hope that war might cease and all nations live together in harmony, he thumped the table with his fist in furious condemnation. Giddy stubbornly refused to concede the moral high ground, and spoke to defend the view that parliamentary reform was a legitimate cause and not in itself seditious, but the spirit of the meeting was against such nice distinctions, and he was obliged to accept the royal proclamation and to lead a unanimous oath of loyalty to it. He was thanked by the meeting, and left it with his authority and dignity both intact, but with a bitter aftertaste of self-betrayal.

Giddy was quick to solicit Beddoes' views, and Beddoes was equally quick to offer them. 'Without delay and without reserve', he wrote back by return of post, 'I think the more you reflect, the more you will repent of having signed that pitiful address'. It had been nothing more than an elaborately staged public humiliation. Giddy's views were already well known, and an open target for gossip and smears; in this situation 'a refusal would not, I imagine, have made your sentiments more known, for defamation is always very busy in disseminating reports concerning religious and political heresy, especially when there is any chance of rendering odious a man who has claims to superiority'. If Giddy, as sheriff, had taken a principled stand, he might have cowed his enemies; but after his capitulation 'they will applaud themselves for having claimed a compleat victory'.[21]

Later in life Giddy would reach precisely the opposite conclusion from Beddoes: that it was 'unquestionably very fortunate that I signed the Address'.[22] He had proved that he could be relied on to square his conscience with his duties, a quality that in time would open the doors of high public office to him. But Beddoes was not thinking of career advancement: this was, for him, a crucial moment for thinking men to express what they felt. He knew that the situation in France was unstable – 'everything seems to indicate the approach of a new revolution'[23] – and that Britain's friends of political reform would need all their courage and eloquence to prevent their nation from becoming a hostage to the uncertain French experiment. He would spend most of the summer of 1792 expressing his views forthrightly in print, and without regard for any royal proclamation. The books and pamphlets he produced would range with extraordinary latitude, advancing his theories not merely on medicine and politics but philosophy, history, religion and education, and in forms as diverse as poetry and polemic, fiction and philosophical treatise. Yet despite their apparent diversity, each would play its role as a sighting shot for his main target which, by the end of the process, would have come clearly into focus.

His first production had in fact already emerged from the presses: a treatise entitled *Extract of a Letter on Early Instruction, Particularly That of the Poor*. It advertised itself as an educational guide 'on the best method of teaching reading and writing',[24] and its ostensible purpose was to argue that rote learning was ill-adapted to the minds of children, and to outline an alternative system 'to bestow that quickness of sight and apprehension, in which ready reading consists'.[25] This was indeed its starting point, but its most remarkable achievement was to extend its analysis of traditional teaching methods into a fundamental critique of British society and politics. Education was a subject that had long been and would remain central to Beddoes' concerns, and that held for him a transformative power equal to, if not deeper than, politics itself.

The transformation that he envisaged had its seemingly modest beginnings in the psychology of learning. Children, he argued, should be taught to write as soon as they are taught to read: in this way they would engage their hands and their minds together, forming parallel associations that would allow cognition and coordination to strengthen one another. At the same time they should be encouraged to study nature, and learn the habits of native plants and animals, which would give them a harmonious sense of their relation to the world. 'Cruelty to animals', he argued, 'is one of the earliest and most pernicious acquisitions of ill-educated children; and yet the same constitution of their nature, disposes them to acquire the habitual sentiment of compassion for both men and animals'.[26] The purpose of education should not be to hammer moral lessons into children's minds, since their moral sense is already in place; rather, it should aim to develop their benign instincts at the expense of their destructive ones.

These were ideas with an immediately recognisable line of descent from the philosophy of John Locke and the writings of Jean-Jacques Rousseau, from which had flowed a stream of primers on liberal education, notably the hugely popular *Sandford and Merton* by the late Lunar associate Thomas Day. But the system that specifically underpinned Beddoes' scheme, and indeed much of his thought, was the psychology of David Hartley, a physician from Yorkshire who, in the 1740s, had developed Locke's theory of associations into a fully articulated model of the brain and its physical workings. For Beddoes, Hartley's system stood alongside John Brown's as a scientific theory of physiology, stripped of medieval and scholastic encumbrances, that offered a solution to many of life's fundamental problems. The fact that Hartley's work would have disappeared had not Joseph Priestley republished it at his own expense was, for Beddoes, 'a proof paramount to all others, of the unconcern of mankind about the true means and ease of happiness'.[27]

Like the Brunonian system, Hartley's was an ambitious extension of Newtonian mechanics into the tissues of life. Where Locke had simply theorised that our ideas arise from our senses and the repeated associations that our perceptions present to us, Hartley had attempted to ground his theory in the physiology of the brain and spinal column. Within these organs, he proposed, were tiny filaments and particles that vibrated as the senses were stimulated; when the same stimuli were repeated, they set up resonances or harmonics to which new sensations attuned themselves. As these vibrations took form, they distilled raw sensation into more complex associations of ideas and stored them as memories, which in turn generated habits of mind and character. It followed from this that education should not be approached as an attempt to force a child's mind into adult patterns of behaviour: it was a far more delicate and fundamental process in which stimulus and sensation could be orchestrated into a harmonious experience of life. The mind of the child was not a passive object to be moulded, but an active one already equipped with the tools to mould itself.

With hindsight Hartley's system, like Brown's, can be seen as an attempt to forge a modern science of body and mind before the necessary discoveries in physiology and neurology had been made. His ideas would continue to be advocated through the nineteenth century by rational philosophers such as John Stuart Mill and, as the anatomy of the nervous system began to give up its secrets, they would eventually be recapitulated to some extent by Ivan Pavlov, William James and the young Sigmund Freud. For Beddoes' generation, though many were sceptical about Hartley's elaborate and speculative architecture of invisible vibrations, particles, ethers and filaments, his vision of the processes of life as physical activity on a microscopic scale offered great explanatory power: Erasmus Darwin, for example, pressed it into service to explain the transfer of information from adult to embryo. But Beddoes' immediate interest in Hartley was more practical, and more revolutionary. If the goal of education is to allow children to develop their characters in their most benign form, then, he insisted, religious instruction must be excluded from it. There is an 'excessive danger', he stressed, 'of strongly attaching to the dogmas of any sect, the minds of those who cannot examine the grounds on which they rest'. To do so is to set up disjunctions between knowledge and belief, to isolate each child within their own sectarian world, at the mercy of the 'oppression . . . that is generally taught for religion'. The prejudices against those with different beliefs, on which the established religions insist, 'brutalise the mind and so entirely pervert our sympathies as to make us feel pleasure from the pain of our fellow-creatures': they are the direct analogue of teaching children to torture animals rather than to empathise with them.[28]

Here, and not in any political experiment in France, were to be found the roots of Britain's current malaise. Indoctrination and intolerance in schools grew naturally into zealotry and the callous exercise of political power, which in turn polluted public life: 'a savage spirit in the people and tyranny in the possessors of power are to one another cause and effect'. To reform politics it will be necessary to 'civilise the people, unless we choose to repose beneath the shadow of bayonets in mercenary and often brutal hands'. The source of Beddoes' imagery is clear: 'and now the Birmingham rioters have presented themselves to my thoughts', he continues, it should be equally clear that a population who had received a liberal and humane education, 'as I wish to see generally adopted among the poor, could never have committed excesses so disgraceful to their age and country'. This was the behaviour of a people who had been brutalised from their first day in the schoolroom; and 'I am utterly unable to conceive how "reading-made-easy", with its endless repetition of godly exclamations, can inspire any human being with benevolence towards his neighbour'.[29]

Beddoes wrote the *Letter on Early Instruction* with his Shropshire constituency in mind: its proposals circulated among the educational community, and had some influence on the style of teaching in local schools such as the Poor House in Shrewsbury. But, as would soon become clear, it was also circulating more widely, and its broader themes were reaching a readership neither expected nor intended.

* * *

Beddoes' next work also engaged with these broader themes, though from a direction and in a style that could hardly have been more different. Over the spring he had continued to exercise his writing hand with verse inspired by Erasmus Darwin; the results had started to take on a life of their own, and he had begun to circulate them among their mutual friends. At a dinner party at William Reynolds' home in Ketley, according to family tradition,[30] a dispute arose over whether Darwin's verse was truly inimitable, to which Beddoes responded by producing a full-blown epic in heroic couplets, abundantly annotated with footnotes and appendices in Darwin's allegedly inimitable style.

Reynolds was so impressed with the work that he paid for the publication of a handsome private edition, accompanied by a set of specially commissioned woodcut illustrations by a local artist inspired by the engraver Thomas Bewick.

Alexander's Expedition down the Hydaspes & the Indus to the Indian Ocean would turn out to be Beddoes' only extensive published excursion into verse. It takes the form of an epic narrative of Alexander the Great's conquest of India: its opening is a florid oration delivered by Alexander, poised on the prow of his fleet's flagship and exhorting his troops to conquest, and its drama opens with a lavish panorama of his army descending from the Indus Valley, its 'cataracts thundering down the shattered steep',[31] where they discover a land of previously unimagined antiquity, inhabited by

> Brahmins old, whom purer eras bore
> Ere Western Science lisped her infant lore.[32]

But this venerable Hindu civilisation is in tragic decline: it is revealed to have suffered for centuries under a series of oppressive tyrants. Alexander believes that destiny has chosen him to write a new chapter in its history, and expresses the hope that his conquest will bring trade and wealth, and

> boundless Commerce mix a cultured world
> from mad mis-rule reclaimed.[33]

But the subsequent narrative reveals the great conqueror's hope to be a delusion, and the poem closes with a climactic vision of a future in which India is once more oppressed under the yoke of Western economic expansion, and

> from Christian strands, the rage accursed of Gain
> . . . pollutes the hallowed shore
> That nursed young Art, and infant science bore.[34]

Darwin's epic confections, with their unintentionally bathetic combination of flowery sentiment and stolid didacticism, have become at the very least an acquired taste for the modern reader, so the appeal of Beddoes' imitation in literary terms may now be beyond recovery. Even its appeal to Beddoes seems not to have stood the test of time, and in the years that followed he would express relief that it had never progressed beyond a small private print run. But the sentiments behind its bombastic couplets would still be easy to recover even if they were not spelled out directly in the extensive footnotes and appendices. The essay that concludes the volume, 'On the Possessions of the British in Hindoostan', regrets that Britain has failed to learn from the humiliating loss of the American colonies that far-flung lands cannot be bullied or oppressed into obedience. Now, with the East India Company building a new empire of corruption and feudal oppression, 'acts of gross injustice' involving 'the ruin and distress of multitudes' are spreading famine across Asia, and flooding Britain with luxuries that, so casually and brutally acquired, feed the avarice and callousness of the British people and corrupt them in return.[35]

Even when following Alexander down the Indus, Beddoes also contrives to turn his tale not only to politics but to chemistry. In an appendix on 'The Complexion of Natives of Hot Countries', he considers the question of dark skin and its prevalence in the tropics, and wonders whether the slow burning of oxygen might be the mechanism that produces it, in the same way that it coats metals with rust. Strong light, he notes, releases oxygen in plants; it also 'turns the combination of nitrous acid and silver black by disengaging oxygen'. Might it release oxygen from human skin and turn it dark in the same way? He has already tested this by experiment, having 'prevailed upon a negro' in Oxford to place his hand in an acid solution, whereupon his fingers 'went white for several days'.[36] Even when he was conjuring distant worlds of deep antiquity, the chemistry of life still forced itself to the front of Beddoes' mind.

But as well as returning to familiar themes, the poem's appendices introduce new dimensions to Beddoes' thought. Most of all, they reveal the depth of his interest in Hindu culture, a newly discovered world of an antiquity far surpassing the timescale allowed by Judaeo-Christian chronology, and with a philosophy containing notions of deep time far more congenial to a Huttonian geologist. He draws the reader's attention to the accuracy of ancient Hindu astronomy, and the fact that it flourished within a religious system that was free of sectarian dogma and oppression (the caste system is dismissed as a later, decadent development). He quotes the *Bhagavad Gita*, which he has been reading carefully for many years, with its great unifying Supreme Being ('I am the Creator of all Things, and all Things proceed from me'); and he urges that the existence and antiquity of such alien creeds should be more widely advertised as an antidote to sectarian strife. 'I know of nothing', he concludes, 'that would so much contribute to soften the heart of blind credulity, and to diffuse peace and goodwill amongst mankind, as a work which should exhibit an impartial comparison of the religious dogmas and morality of different nations'.[37]

* * *

Following *Alexander* off the presses was a very different work: a lengthy treatise on theories of knowledge entitled *On the Nature of Demonstrative Evidence*. Beddoes typically had little patience for abstract reasoning, and his tenacious engagement with it at this point reveals the extent to which he regarded the definitions of evidence and proof as an urgent practical problem. The route he was about to travel could, he recognised, only be validated by experiment and evidence, and his assumption that this process afforded a superior route to knowledge than tradition and revealed truth was one that needed to be spelled out and justified.

Experiment, and the status of the facts that it generated, could be construed in various different ways. In the Aristotelian tradition – as, for example, in most university teaching – experiment tended to be used as a rhetorical form of demonstration, an illustration of an accepted theory rather than a strict proof of it. To Beddoes, however, it meant something quite different: for him, as for his touchstone Francis Bacon, an experiment was the acid test of a hypothesis, which it had the power to confirm or deny. It was, however, not possible for experiment to stand entirely alone: no experiment could produce truth without a theory against which to test its results. The facts that it established, even if perfectly valid in themselves, might show a pattern that was illusory, an artefact of the process of investigation itself: the experimenter must always be vigilant against findings that reflected his own focus and preoccupations rather than an external reality. If evidence produced by experiment was to be accepted, the ground rules of its relationship to theory, and the wider questions of how we observe the world, needed to be clearly established in advance.

Beddoes' contention, developed along various parallel tracks, was that we perceive the world only through our sensations, and that all knowledge must ultimately be testable against them. Even mathematics, typically cited as a perfect abstract system impervious to material evidence, had from its beginnings in Euclid been experimental: geometry was not an illustration of a pre-existing theory of the world, but a set of conventions around which our view of the world could be configured. Beddoes highlights the particular problems that pneumatic chemistry faces in this respect: because the airs are imperceptible to our senses they may seem imponderable, and the theory of their actions vaporous and obscure, but anyone who has understood Lavoisier's work will recognise that his conclusions are, in fact, as solid 'as any proposition in Euclid'.[38]

There were, naturally, lessons here for education: experiment and demonstration remained the best way of teaching geometry to children, and mathematics, abstract as it might appear, should 'strengthen, if possible, those arguments which have been used in favour of a plan of education which shall pay some attention to the senses'.[39] But there were also important questions for anyone about to embark on a project that would use experiment as a tool to reconfigure the practice of medicine. How would Beddoes be able to convince himself and others that he was not simply gathering evidence that squared with his theory, and ignoring evidence that contradicted it? And were sick patients proper subjects for experiment in the first place, especially by one who had taken the Hippocratic Oath?

On the Nature of Demonstrative Evidence was dedicated to Davies Giddy, to whom his former teacher already deferred on the subject of mathematics, and Beddoes signed the preface from Oxford: this was a subject for which his university credentials were still a valuable badge of authority. But the most striking name within its pages is that of 'Mr. Kant, Professor at Koenigsberg in Prussia', whose philosophy is laid out here in English probably for the first time. 'Mr. Kant asserts that we are in possession of knowledge *a priori*', Beddoes informs the reader, before including a substantial passage from the previously untranslated *Critique of Pure Reason*. On this first encounter Beddoes is briskly dismissive: 'This passage, as I apprehend it, includes a considerable number of mistakes, some more, some less connected with the subject of mathematical reasoning'.[40]

Kant's search for *a priori* truths, absolute and independent of sensory evidence, appears to Beddoes at first sight to be pointless: he is happy to follow Locke and Hartley in accepting that all knowledge is ultimately provisional, that reality is simply the best approximation that our senses can make of the world, and that even mathematical theorems are simply observations that turn out repeatedly to be true. But Kant's curious-looking assertions about the possibility of *a priori* knowledge were more than, as Beddoes first assumed, the usual attempt to defend religious authority from the light of reason: Kant was not attempting to dethrone science, but to root its conclusions more deeply. Newton, after all, did not invent gravity, nor did he observe it: our perception of its effects is enough for us to know it is real. But if reality is beyond our ken, what precisely was it that Newton's experiments had observed, and his laws so accurately predicted?

Kant would later admire Beddoes' writings on consumption, but Beddoes would never learn to read Kant with pleasure: in a later review, he would praise his 'universally interesting' subjects but lament that he was 'so utter a stranger to clearness'.[41] He would nevertheless become, both directly and indirectly, a prime conduit for the new German philosophy's influence on British thought.

*   *   *

Alongside these excursions into outlying territory, Beddoes was working over the summer on two publications directly focused on launching his medical career and researches. Both would be extremely successful. The first was addressed neither to the medical profession nor to the scientifically minded public but, most unusually, to the labouring classes of Shropshire. *The History of Isaac Jenkins* was a parable of sickness and health, presented in the form of a short, melodramatic but vigorously descriptive novel, convincingly set in the milieu of rural Shropshire in the infamous autumn of 1783. 'The blustering winds and the pelting showers went on all autumn long', he began, in the knowledge that his readership would recall all too well how the weather had 'battered the wheat-lands and made the clays as stiff as if they had been trodden on purpose to make bricks'. As crops failed and famine encroached, 'there came a great sickness all over the country; and numbers died of the spotted fever, especially among the poor'.[42]

Into this grimly familiar scene Beddoes introduces 'Isaac Jenkins, a poor labouring man', and his wife, Sarah, whose children are infected by the spotted fever.[43] He paints a bleak picture of the medical options at a poor mother's disposal: a choice between a doctor who 'lived at a distance, quite at Ludlow; and she could not pay him for his physic, much less for his journey', and a 'quack-doctor' who had 'left some white powder that was nothing but salt-petre . . . for your quack-doctors care not a farthing whether they kill or cure; all they want is to fleece those that know no better'.[44] Isaac's response to his family's impossible situation is to turn to drink, a habit that began after he and Sarah had lost a young son to a tragic accident in which, with resonances of Beddoes' grandfather, he was trampled by a horse and left 'bleeding and mangled with his face all one wound'.[45] Isaac spends the last of the family's money in Big Martha's alehouse, returning home drunk and insensible to the misery and suffering he is inflicting on his loved ones.

The situation is rescued by the *deus ex machina* of a passing doctor who explains to Isaac the tragic consequences of drunkenness and restores him to the early days of his marriage when 'his house was always clean and the children ruddy and plump . . . not pale in the cheek, and pot-bellied, as if they had the worms'. Isaac is restored, having 'felt too sensibly the difference between beggary with drunkenness and discontent, and plenty with sobriety and a light, cheerful heart'.[46] Although the ending is mawkish and didactic, for most of its length Beddoes' story reflects its world with an unsentimental realism to which its intended readers responded enthusiastically. Shortly after its publication, he announced proudly to Giddy that 'it is a prodigious favourite in this part of England', already into a second edition with nearly five thousand copies sold. 'Among the colliers', he reported, 'it is a common saying when a man is seen staggering along – he has been at Big Martha's', and indeed 'many gentle & all simple folks' seemed to believe that the Jenkins family were real people.[47]

Isaac Jenkins was printed in a small chapbook-style format and sold cheaply, and its success among Shropshire's rural poor was largely due to a network of which Beddoes strongly disapproved. Its leading light was the poet, dramatist and fervent religious convert Hannah More, who had set up an imprint called The Cheap Repository Tracts to provide improving works of advice to the poor. The enterprise was generously funded by private subscription; its most serious limitation was that the poor, well aware they were being sermonised to, showed little interest even in free copies of the tracts on offer. More's own works, whose titles included *Practical Piety* and *Christian Morals*, represented everything Beddoes wished to eradicate from education – they were, in his view, 'calculated to fill the minds of the uneducated with superstition & bigotry & hypocrisy'[48] – and he initially declined to have his work distributed on her list. But Hannah More was also, along with her friend William Wilberforce, a committed agitator for the abolition of slavery, and eventually Beddoes relented and made common cause with her. He had produced a rare commodity, a work that her intended readers did not feel patronised by and that entertained while it informed, and *Isaac Jenkins* remained in print through her religious tract society until the 1840s. Despite Beddoes' concerns that it might not 'find favour in distant parts' because of Isaac's thick Shropshire dialect and 'other local peculiarities', it sold widely across the entire country.[49] It would give him a distinctive national profile as a physician who could speak with authority to the poor, and in some literary quarters it remained his best-known work: Samuel Taylor Coleridge would refer to 'the genius of him who wrote *Isaac Jenkins*'.[50]

* * *

Beddoes' second medical book was aimed squarely at the profession itself, and it was his most significant product of the summer of 1792: a work with a highly ambitious sweep that aimed to make his pneumatic medical project seem not only comprehensible but urgently necessary. His strategy was three-pronged: to begin in territory that was of universal interest to physicians, by offering practical therapies for common medical conditions; to advance smoothly and persuasively towards a distinctive theory that might account for their efficacy;

and finally to prepare his readers for a trial of his theory by an unprecedented programme of experiment. Its dedication was signed from Oxford on 30 July 1792, although he was almost certainly in Shropshire at the time: this was another occasion for highlighting his university credentials. Published in January 1793 under the title *Observations on the Nature and Cure of Calculus, Sea Scurvy, Consumption, Catarrh and Fever*, it offered novel and empirically tested remedies for all these conditions that combined to form a manifesto for a new chemical medicine that was, he insisted, 'daily unfolding the profoundest secrets of nature'[51] and would soon transform the job of the physician entirely. These remedies were not to be confused with the apothecaries' fads to which medicine had long been subject – understandably so, since 'the physician stalks abroad with greater dignity when he feels a full quiver at his shoulders, however blunt the arrows it contains'.[52] These were, rather, the products of 'a boundless region of discovery ... opening up before us', and which, he announces, characteristically unable to resist the comparison, 'will effect a greater improvement in the morals of mankind, than all the sermons that ever have been, or ever will be preached'.[53]

Beddoes proceeds by examining each of the conditions in the title in the same empirical manner: an analysis of their symptoms and a theory that accounts for them, buttressed by his characteristically visceral descriptions and a wide selection of case histories, both observed by him directly and collected from other physicians. But in each case he is particularly attentive to the relationship of air, and specifically oxygen, to the familiar pathology. In consumption he notes, as had many before him, that 'the cheeks appear as if painted with a circumscribed spot of pure florid red; the lips and tubercles in the canthus [corners] of the eyes are also redder than in health'; and he asks the reader to consider that this indicates an excess of oxygen in the blood.[54] From this might follow various possibilities: it could be that the tubercular inflammation causes the blood to absorb more oxygen from the lungs or, conversely, that hyperoxygenated blood causes the inflammation. It may also be significant that this stage of the disease tends to recede during pregnancy, when the oxygen in the blood is being shared with the foetus. But here Beddoes reaches the limits of experiment: 'a comparison of the arterial blood of phthisical [consumptive] and healthy persons', he sighs, 'would be so very interesting on this account, that I wish it were practicable'.[55]

As he follows his thread through the seemingly disparate diseases of his title, he edges towards an overarching theory that might yoke them all together. Catarrh, like consumption, he classifies as a disease of excess oxygen; others, including obesity and scurvy, as diseases of an inability to absorb enough of it. In the case of scurvy, for example, the opposite effect to consumption is clearly visible: the 'gradual abstraction of oxygen from the whole system, just as death is produced in drowning'. The blood becomes pale, the patient wan, and 'large livid spots' appear as colour ebbs from the body.[56] Although the naval physician Thomas Trotter had recently proposed that scurvy was caused by the lack of fresh vegetables, Beddoes remained unconvinced: what, for example, of the Laplanders, who survive their winters exclusively on raw meat and never develop the condition? 'Captain Cook's unexampled success in preserving his crews from the scurvy during his last two voyages' was, in Beddoes' opinion, more plausibly attributed to his 'extreme care to keep his ships well aired' than to his supply of fresh limes: the crew members without fresh food were also the ones confined in cramped, airless and poorly oxygenated conditions below decks, and were equally those, along with slaves, who had the highest mortality rates.[57]

Beddoes was advancing a bracingly novel approach to medical treatment: one that submitted all its assumptions to experiment and evidence, and with which he aimed to shine the light of science into a profession that had thus far derived distressingly few practical benefits from it. But he was well aware that the idea of an experimentally based medicine was still problematic. The term 'empirical' had yet to acquire its modern sense of scientific rigour; it was, rather, a common term for an impostor or charlatan, a virtual synonym for 'quack'. It had its origins in an ancient Greek school of physicians who had practised medicine on the basis of experience, and who had been mocked by theorists such as Aristotle for their bumbling reliance on trial and error. Since the Middle Ages 'empiric' had come to acquire the more specific meaning of a doctor without university qualifications, and by extension a pretender for whom the doctor's garb was simply a means of extracting money from his patients.

Beddoes despised empirics as heartily as did any bewigged and buckled physician of the old school; perhaps more so, as he saw their opportunism not just as a slur on the medical profession as it was currently constituted, but as a serious obstacle to its reform. He was stung by those who confused his empirical method with quackery, and constantly alert to the need to rebut them: he would habitually include quotations from Francis Bacon on his title-pages to make it clear that his goal was not simply to peddle lucrative nostrums but to advance systematically by tests and trials towards a new understanding of disease and its treatment. In this sense his oxygen theory – along with his university degrees – was a rhetorical defence against those who attempted to tar him with the empiric brush. But appealing to theory brought its own dangers, of which his conclusions about scurvy were a good example. His evidence concerning the Laplanders was culled from an impeccable source – the travel journals of the great Swedish naturalist Carl Linnaeus – but he gave it prominence over other evidence because the efficacy of fresh vegetables argued against the overarching oxygen theory that he was assembling. He was guilty here of being not too empirical, but not empirical enough to challenge sufficiently the evidence that favoured his conclusions.

Observations, however, had more than a theory to sell: it also had a project, the form of which became more distinct as the book progressed. 'The more you reflect', he suggests to the reader, 'the more you will be convinced that nothing would so much contribute to rescue the art of medicine from its present helpless condition, as the discovery of the means of regulating the constitution of the atmosphere'.[58] If, for example, consumption is a condition of hyperoxygenation, might it not be treated by airs poor in oxygen, and rich instead in hydrogen or nitrogen? If kidney stones and gallstones are alkaline deposits, might they not be dissolved by acidic carbon dioxide? The idea of a pneumatic research institution is yet to be formally proposed, but the medical world is being softened up for its appearance on the stage, and warned of its revolutionary implications. The chemical sciences 'are fast conducting us to a more intimate knowledge of ourselves', and it is to be expected that they will, sooner or later, generate more advanced forms of treatment.[59] Beddoes' project may offer more arrows for the physician's quiver; but in the long term it may equally make him, in his present form at any rate, redundant.

As the pneumatic research project begins to edge into the frame, it brings with it another potential objection to be confronted: Beddoes might now be seen not only as an empiric but also as a 'projector', a term still coloured with its Jacobean sense of an alchemist claiming the power to transmute base metal into gold, as well as the broader meaning of an entrepreneur whose claims to scientific authority are, in reality, part of his sales pitch. Among the quacks, there were many with their own brands of snake oil, advertised with glowing patient testimonials and the trappings of plausible-sounding medical theory; many, too, had eye-catching 'experiments' that seemed to demonstrate miracle cures, but were in reality no more than gimmicks designed to convince the gullible. As such mountebanks demonstrated, it was not hard to assemble a superficially impressive body of testimony and case history: one had only to showcase the positive evidence, and ignore or suppress the negative. As Beddoes moved towards establishing his project, he would need to become an advocate for his own cause without deluding himself or betraying his Baconian principles.

There was a further obstacle that Beddoes faced as his theory came into focus: its apparent debt to the Brunonian system. In both, every disease was caused by either too much or too little of one fundamental principle – excitation in Brown's theory, oxygen in Beddoes' – and in both cases the prescription was a choice of one or another chemical remedy. Given that most of the medical profession saw Brown as little better than a quack, it was important to Beddoes that his own work was not received as a speculative refinement of a system widely held to have been discredited. His chosen line of defence was not to deny the connection but to assemble a body of authority in support of Brown's theories, notably 'the author of the *Botanic Garden* [i.e., Darwin], who is no less eminent as a physician than as a poet',[60] and the German physician Christoph Girtanner, an old Edinburgh colleague, from whom Beddoes appended a paper identifying oxygen in the blood as its animating force, and arguing that Brown's theory made more sense within the new chemistry of Lavoisier than it had within the outmoded phlogistonian system. Girtanner was now at the University of Göttingen, where 'der Brownismus' had won considerable support from the medical profession, and his inclusion strengthened the impression that Beddoes' case was supported by experts not only in Britain but in other nations whose languages Beddoes, unlike the vast majority of his readers, had mastered.

This was a strategy that Beddoes would continue to employ throughout the impassioned public campaign that *Observations* had begun, and that would build in intensity as the decade progressed. His books would habitually include contributions from others, particularly the case notes of doctors whose experiences and observations mirrored his own, but also the results of experiments that chimed with his theories. Partly, this would be because the scope of his project would expand beyond his abilities to realise it alone; partly, it would reflect a generous instinct to share the limelight with his colleagues and protégés; but it would also address the need to present himself not as an individual but as a movement, and to represent his idiosyncratic project as the work of a concerted team of experts. It was a strategy that would throw a double image over his work to come: both of an embattled individual swimming against the current, and of a figurehead presiding over a teeming scientific movement. But it would also create confusion, not only in the minds of others but also in his own, about what precisely was being asserted by the author as fact, and what was merely the provisional, suggestive or speculative observations of others.

* * *

Beddoes' writings over the summer of 1792 had covered almost every subject imaginable with the conspicuous exceptions of chemical theory and geology, the subjects with which he was engaged at Oxford. He had made great strides towards establishing a new career: he had learned to address himself to new readerships – provincial industrialists, practising physicians, the rural poor – and to present himself not merely as a scholar but as a professional physician and popular author. At thirty-two he was finally making the transition for which he had been so restless, from eternal student to man of the world. His new career would free him from financial dependence on his family, perhaps to start a family of his own. For all these reasons his disengagement from Oxford was perhaps inevitable; but it would not proceed as planned, and would saddle him with a reputation he had not anticipated.

The application for a regius chair in chemistry, which had looked promising when Beddoes left Oxford in June, had run into difficulties. Once Beddoes' name had been put forward, Lord Dundas' Home Office had begun to make checks on his character, and these had immediately rung loud bells of alarm. Dundas received a report from his Oxford source that, although Beddoes was a fine chemist and his lectures exceptionally well attended, 'in his political character he is a most violent democrat' who 'takes great pains to seduce young men to the same political principles'.[61] Beddoes' name also appeared alongside that of Joseph Priestley on a Home Office list of undesirables, one of many that had begun to circulate since the royal proclamation had called on local authorities to expose political dissenters. Once Dundas was informed that Beddoes had already privately offered his resignation, it was clear that the application for the chair would go no further in its present form, though it might be reopened once a new chemical reader was in post.

The outcome seemed likely to be the one that Beddoes had proposed at the beginning of the summer: he would complete his next set of lectures and then resign, leaving the pursuit of the regius chair to the next incumbent. But this, too, was a solution that was not to hold. Beddoes had one more pamphlet to publish before 1792 was out; and it was one for which he would be long remembered.

By September, even allowing for the slant of the British press, it was clear to Beddoes that the new revolution he had foreseen was poised to tear the young French republic apart. The Assembly was riven between factions, Girondin, Jacobin and Dantonist; rumours were building that a monarchist *putsch* was simply awaiting its moment; Jean-Paul Marat's newspaper *L'Ami du peuple* was calling for the blood of the traitors to flow, and an inflamed mob had stormed the Tuileries Palace and butchered Louis XVI's Swiss Guard. Beddoes professed himself, not for the first or the last time, entirely sick of the tawdry drama. 'Now that the National Assembly by their incapacity & cowardice have united everything that is ridiculous with everything that is horrible', he wrote to Giddy, 'one has not the smallest inclination left to take the trouble of thinking about France'.[62]

He was, for this reason, especially delighted to discover 'a diversion of thought at a moment when one is so much inclined to be dissatisfied with everything & every person, not excepting one's own dear self': he received the first reports from Italy of Luigi Galvani's discovery that the flayed muscles of a dead frog will twitch when connected with wires coated in lead and silver. This had sent Beddoes into a frenzy of experiment, dissecting frogs and mice to duplicate Galvani's work and rolling pieces of lead and silver around on his tongue to taste 'the electrical aura' that they emanated. It was a discovery that boosted his long-standing conviction that the secret of muscular action was on the verge of being unlocked, and that the fundamental motive force of life, thus far only guessed at by Brown's excitation and Hartley's vibrations, was on the point of being made manifest. 'In a twelvemonth the face of medicine will be changed', he predicted: once science had tracked the life force to the tissues and fibres, 'there will soon be an end of physicians and apothecaries; or at least a gradual diminution of a body of men, who do and have done infinitely more mischief than good'.[63]

But within days, when news arived that the Paris mob had broken into the city's jails and butchered the royalist sympathisers they found there, Beddoes' resolve to ignore the French experiment crumbled. *The Times* reported the sacrifice of hundreds of royalist prisoners to 'the brutal fury of the mob', and concluded with the deepest solemnity it could muster: 'Read this, ye Englishmen, and ardently pray that your happy Constitution may never be outraged by the despotic tyranny of Equalisation'.[64] Although 'for obvious reasons' *The Times* held back on distressing details, other newspapers were full of pikings, beheadings, carts piled high with mutilated bodies and drunken mobs feasting on roasted human genitals.

Beddoes, returning to the news from a medical visit to Wales, found the press reports of the September Massacres extremely suspicious. It had been months since any British newspapers had had their own correspondents in Paris; this news was being propagated by a network of aristocratic émigrés with a vested interest in marshalling British support for a royalist counter-revolution, and amplified by British newspapers seeking to outdo one another with lurid claims of mass murder and cannibalism. Despite the confusion, however, it was abundantly clear who the beneficiaries of such reports were: all along his route through Shropshire, market stalls were stacked with pamphlets from a society raising funds 'for the relief of the suffering French Clergy refugees in the British dominions', and citing the 'horrid barbarities that have raged with unexampled violence' in France. The more extreme the reports of the massacres, the better for the campaign to establish France's Catholic clergy in exile on British soil.[65]

Beddoes' response was a double-sided handbill, printed in Shifnal on 9 October and entitled *Reasons for Believing the Friends of Liberty in France Not to Be the Authors or Abettors of the Crimes Committed in that Country*. These claims of atrocities, he began, made it more important than ever to humanise the French people – 'the most injured and most enlightened people on earth'[66] – who, despite having 'improved science more than all other nations put together', were now being caricatured as bloodthirsty barbarians. For all the 'calumny that has been vomited forth against the French reformers', their revolution had already achieved much. Were the British to read the French press, they would learn of an end to tithes, 'that accursed relic of popery', and to the feudal taxes on the poor; of agriculture 'wonderfully improved', 'prompt and cheap justice' and the prospect, if only the rest of the world would allow it, of 'an honest, cheap and unoppressive government'. But they were being told none of this; rather, they were being fed 'fabricated lies', and 'in this infamous manner are Englishmen mocked: by whom?'

Beddoes' answer to his own question was that it was precisely those begging to be rescued from the massacres who were to blame for them. Such spectacles of animal brutality are 'exhibited only by nations first brutalised and then instigated to fanatical fury by the Priesthood'; indeed, it can be observed universally that 'the less ecclesiastical influence, the less of deadly animosity among men'. The grim violence of the French Revolution has its roots in a poisonous clerisy, 'stung to madness at being no longer able to keep a mighty nation yoked to the car of imposture', and making 'a last effort to restore despotism and superstition'. The just and reasonable aims of the Assembly had been thwarted by a king who had made no secret of his hopes 'to be exalted again to arbitrary power by the engine of Priestcraft'. Once the king's cynical manoeuvres were exposed for all to see, it could be no surprise that 'these pests of society were exiled', and 'this is the true reason why we see them in England'.

Though deeply sceptical of the British press reports, Beddoes stopped short of denying that the massacres had taken place, or that they had escalated the violence of the revolution to new heights. 'The mind of the French nation', he insisted, 'has been deeply wounded by the enormities of September 2nd'. They too feel horror and shame at the crimes of that day, and 'are now acting as they ought in order to prevent their return'. But these brutal scenes cannot simply be laid at the door of political reform; or how could we have seen, as we did last summer in Birmingham, 'the design to roast Dr Priestley alive', the expression under the banner of Church and King of 'as sanguinary a spirit as the Paris mob'? And precisely as it would not be 'justifiable to criminate the whole English nation on this account',[67] so we must view the appalling scenes in the Paris streets with enough clarity and courage to recognise the very same spectacle that we witnessed in Birmingham last Bastille Day: a people driven to bigotry and hatred by the divisive machinations of the church.

In the heat of the moment, Beddoes had released his frustrations at the injustice of the British press towards the French experiment in a cathartic blast; but he had chosen the worst possible time to do so, and particularly to argue for a moral equivalence between the Paris massacres and the Birmingham mob. The September Massacres were, across the length and breadth of Britain, the trigger for precisely the opposite response, the final proof that the French Revolution represented not liberty but the greatest threat to it imaginable. As a result, Beddoes' polemic found its way with a far greater rapidity than his previous works into the hands of those who moved in circles very different from his intended readers in liberal and dissenting Shropshire. The combination of Oxford and Shropshire made him more conspicuous than he might otherwise have been: in Oxford, an establishment figure who was being investigated in the course of his application for a politically sensitive post, and in Shropshire, part of a civil society in the habit of expressing its political opinions freely, and cushioned from the panics that were sweeping the rest of the country.

Beddoes had done much to manage a graceful exit from his previous life, but this handbill had left him badly exposed. Now the Home Office, stepping up its efforts to track 'British Jacobins', identified him as one of that faction. Evan Nepean, the under-secretary who was coordinating the Home Office's rapidly expanding network of spies and informers, wrote to the Shropshire MP Isaac Hawkins Browne, a landowner and ironmaster who lived close to Reynolds, to alert him to the fact that 'Dr Beddowes [sic] has lately been very active in sowing sedition in your neighbourhood, particularly by the distribution of pamphlets of a very mischievous and inflammatory tendency'[68] – the plural suggesting, perhaps, that his other writings over the summer were also under scrutiny.

By the beginning of November, Beddoes' disgust with events in France had already returned, and he was complaining to Giddy once more about the 'infernal' influence of Marat and Robespierre's faction on the Convention.[69] But by this point he was aware that he had become 'eminently and much beyond my importance, odious to Pitt and his gang, as I know from an hundred curious facts'.[70] He had fallen into precisely the trap from which Giddy, however invidiously, had been protected: he had become identified as a Jacobin, a smear that would stay with him for the rest of his life and would remain a badge of pride only in dwindling circles. While others would reposition themselves discreetly as their youthful ardour faded, Beddoes would always be marked by militant alienation from the establishment, a stain that would never be wiped clean.

The effect in Oxford was to hurry his resignation. Although he had offered to continue teaching into the summer of 1793, he was advised to resign at Christmas: gossip had spread about his politics, which meant that students would give him a wide berth and he 'should have no class'. A successor was appointed to the post of chemical reader, and Beddoes' send-off was cordial: except for one colleague 'who would not speak to me, I have remarked an unusual forwardness of civility in the rest of my acquaintances', a fact he attributed optimistically to 'the increased liberality of the age'.[71] But this was not how his departure would be remembered by others: throughout the rest of his life and beyond, he would be followed by the story that he was forced to leave Oxford on account of his political views. While not strictly accurate – technically, he had resigned before politics had become a defining issue – it was a story that contained enough truth to endure.

*   *   *

But if Beddoes had been reckless in some respects, in others he had prepared his ground carefully and, by the end of 1792, as one door slammed shut, another was opening. He returned to stay with Reynolds in Ketley over Christmas and, in his laboratory, took his first practical steps in pneumatic medicine. His first patient was the son of a Mr Crump, a surgeon-apothecary in the nearby town of Albrighton, who had developed a cough that had run to a chest infection and fever. Beddoes, Sadler and Reynolds had rigged a tube that diverted gas from a reaction in a laboratory into a breathing apparatus beside the bed, and the Crump boy was treated with 'a mixture of airs'.[72] Despite rudimentary equipment that made the airs hard to administer with accuracy, some relief of the boy's symptoms was immediately apparent. His cough subsided, his expectoration was relieved, the foulness in his breath vanished and he found himself breathing more easily.

Beddoes now advanced to a more serious trial, using the most appropriate experimental subject available: himself. His first instinct, as it had been with Galvani's lead and silver wires, was to test the effects directly on his own person. He was not unusual in choosing to use his own body as a living laboratory.

Many of the great breakthroughs in medicine and anatomy had been made in this way: Lazzaro Spallanzani, for example, the great Italian physiologist whose work Beddoes had translated, had established the digestive power of gastric juices by swallowing and regurgitating linen bags. Self-experimentation also relieved Beddoes of the ethical responsibility of experimenting on the sick, and left him answerable to no one but himself. Despite his chronic wheezy chest, he had no history of consumption, and felt reasonably safe in exploring the effect of a greatly increased supply of oxygen on his own system.

He began by inhaling the gas for four or five minutes at a time, gradually increasing the frequency of his sessions until he was taking it for a total of an hour a day. While breathing it, he felt 'that agreeable glow and lightness, which has been described by Doctor Priestley and others'; but as the doses built up, he became aware of deeper and more persistent systemic effects. He experienced 'a much greater flow of spirits than formerly', a firmer muscular tone, and an increased floridity in his complexion, with a marked 'carnation tint at the ends of the fingers'. He was, he confessed, 'rather fat' when the experiment began, but during the course of it he 'fell away rapidly, my waistcoats becoming much too large for me', despite indulging an appetite greater than usual. And even in the biting easterly wind that prevailed, 'I never once experienced the sensation of chillness'.

Feeling nothing but benefits, he persisted, but soon started to experience less welcome effects. His skin began to dry out, he felt his pulse race and hot flushes course through his system; during a long journey by mail coach and on horseback he became 'flushed and hectic' with fever.[73] He began to have regular nosebleeds, something from which he did not usually suffer, and the blood that stained his handkerchief was abnormally bright. The same bright blood flowed more easily from cuts, and was harder to stanch. Returning to Ketley, he took to his bed and recuperated under the care of Reynolds and Dr Yonge, the Shifnal surgeon who had encouraged his medical career ever since their shared vigil at his grandfather's deathbed. Beddoes recovered swiftly, and with high spirits. His trial had demonstrated emphatically that oxygen, taken to excess, had the capacity to duplicate many of the distinctive symptoms of consumption.

By the early weeks of 1793 the idea of a medical pneumatic institution had become a practical proposition, and Beddoes did not have to look far for the funds to develop it. William Reynolds offered to support it with a donation of £200, a figure that his brother Joseph and Dr Yonge both agreed to match. Beddoes was able to contribute the same amount himself, presumably from his family inheritance.[74] This seed money, the equivalent perhaps of £50,000 today, was enough for him to begin searching for a location, and establishing a laboratory, which Joseph Sadler offered to assist in constructing.

At least as valuable as the cash was the endorsement of Erasmus Darwin, which arrived in the form of a glowing response to a proof copy of *Observations on the Nature and Cure of Calculus*. Darwin declared himself delighted that Beddoes was proposing to do battle wih consumption, 'this giant-malady, which has hitherto baffled the skill, and withstood the prowess of all ages'. He agreed that 'a variety of experiments with a mixture of airs' held the promise of dramatic new insights into the condition, and also eminently practicable avenues for treatment, since it would allow patients to keep up other prescriptions of medicine or diet 'without interfering with your remedy which is immediately applied to the seat of the disease, and the ulcerations of the lungs'. 'Go on, dear Sir', Darwin concluded with genial magnanimity, 'save the young and the fair of the rising generation from premature death; and rescue the science of medicine from its greatest opprobrium'.[75]

Beddoes' year of hiatus, with its restlessness and recklessness, was at an end; his project was set fair and ready for launch. But he was still unsure where to establish it. By its nature, it required a peculiar set of conditions in which to flourish. It needed the support of a community of wealthy patients and benefactors, and a base from which to run a subscription list for financial supporters and communicate findings through the networks of the professional and the influential. Beddoes would need to fund himself, and the project, by working as a physician, with the wealthy among his clientele. But he would also need a community of poor patients experiencing various stages of a spectrum of diseases, particularly consumption. The research would require a steady stream of experimental subjects, a very unusual requirement and one of which many patients would be suspicious: his best strategy for obtaining them would be to offer free treatment in exchange for participation in his experimental work to a class of patients who could otherwise not afford a physician. It was, in any case, by the treatment of the poor that the project would stand or fall. There was no shortage of ingenious and expensive treatments available for the wealthy, who could afford to travel for their health: spas and resorts across the world were growing fat on their beneficial airs. It was the rich who would pay for the project, but the poor who must benefit.

Beddoes went to London with James Sadler and Dr Yonge in tow to search for an appropriate site, but their initial impressions were unpromising. London was Oxford writ large, its medical community traditional and staunchly royalist, and its focus exclusively on the wealthy; its poor districts were another world, and an institution set among them would starve just as they did. As Beddoes began to cast around for alternatives, Darwin suggested that he should divert their search to Bristol. It was cheaper by far than London, yet still a busy and bustling city, and, like Shropshire, with a wealthy network of dissenters and merchants that intersected in many ways with the Lunar Society and its associates in the Midlands. The centre of the city, around the docks, teemed with the labouring poor; and, on the outskirts of Bristol itself, the outlying spa of Hotwells in the Avon Gorge attracted a stready stream of consumptive patients. Perched above Hotwells was Clifton, one of the most beautiful villages in the country and an ideal place of residence.

It was a shrewd recommendation. Beddoes, though he had never lived in Bristol before, would take up permanent residence. In Bristol, and its satellites of Clifton and Hotwells, he would form a scientific and social circle unlike any the city had witnessed before, and that would bring his project to fruition in ways he could never have foreseen. And within weeks of relocating there he would have met not only the man who was to become his main financial backer, but the woman who was shortly to become his wife.

3

The Projector

By the early spring of 1793, as Beddoes was packing up his life in Oxford and preparing his move to Bristol, Britain was a nation at war. The transition from sabre-rattling and whispers of sedition to all-out hostilities had taken place with a rapidity that had astonished and appalled him. He had judged the Assembly's narrow vote to execute King Louis XVI for treason as inevitable, 'a measure of salutary justice';[1] but he had not anticipated the near-inevitability that Britain would take the act as a declaration of war on monarchy itself. He had been profoundly moved by the success of the French revolutionary under-dogs against the might of the Austrian war machine – 'I am not acquainted with such an effort of virtue shown by such & so large a body of men under every species of discouragement'[2] – and was horrified that Britain was now joining battle on the side of the oppressors. Ever since the Glorious Revolution the despotic and unreconstructed monarchies of Austria and Prussia had been held up as examples of the tyranny from which Britain had freed herself; now, they were to be her allies in snuffing out the flame of liberty.

With the declaration of war, ports on both sides of the Channel had been blockaded, which had in turn choked off Bristol's commercial lifeblood. 'The buildings here at Clifton exhibit the most melancholy appearance', Beddoes wrote to Giddy soon after his arrival. 'Five weeks ago half the masons and carpenters were discharged; and three weeks ago all the remainder – I see not a single labourer at work – not one house in ten is covered in and I am told there is not the smallest chance of their being so before winter – part of the men enlisted – part starving.' Clifton was an ominous bellwether for the state of the nation. With a thriving mercantile city to one side and some of the most pictur-esque scenery in Britain to the other, it had become a highly desirable address, and wealthy Bristol traders were surrounding its traditional cottages with the tawny sandstone villas and crescents that had recently transformed nearby Bath into a byword for elegant modern living. Now, the frenzy of sawing and hammering was silent, and the residents were looking anxiously to Bristol's docks and wondering how long the warehouse stocks would last.

From his vantage point in Clifton, Beddoes could see the shockwaves of recession rippling across the country. Bristol was a hub not only of the Atlantic trade from the West Indies and Newfoundland, but of the river traffic that flowed through the Bristol Channel, up the Severn and through the Midlands, where it diffused through the network of canals that the likes of Josiah Wedgwood and Matthew Boulton had done so much to foster. But this network was also grinding to a halt. 'The great Worcester canal began with great spirit three months ago', Beddoes told Giddy, but 'three weeks ago 400 hands were discharged, and I am told the rest have been paid off since'. All of William Reynolds' large orders for pig iron had been cancelled, and he was being forced to lay off five hundred ironworkers at Coalbrookdale. 'The poor men have been remonstrating', he confided, 'they ask why should I be discharged rather than another, and then mutter an oath that they would not starve'.[3]

This was an oath that Davies Giddy had heard frequently in Cornwall over the past weeks. In January, while staying in the fishing port of Looe, he had been obliged to call out the county militia for the first time to confront a mob of three hundred miners who had been laid off from their jobs and were advancing on the harbour, where they had heard rumours that grain was being hoarded. Giddy had been open and reasonable with them: he listened to their grievances, opened the warehouses to show how low stocks had fallen, and gave each of them a shilling to buy food for themselves and their families. But three weeks later they had returned, hungry and angry, and this time Giddy had been forced to order the militia to disperse them with a volley of grapeshot. In private he maintained his principled refusal to take sides, but in practice his convictions were being rendered irrelevant.

In Bristol the war and its consequent economic squeeze had exacerbated social divisions that ran as deep as anywhere in the country. For most of the century it had been England's undisputed second city, a boom town fuelled by a trade that encompassed everything from Spanish wine to Newfoundland fish, but of which by far the most profitable was the 'triangular trade' with Africa and the Caribbean in sugar and slaves. It had become hugely prosperous, giving rise to a popular stereotype of a smug, *nouveau-riche* business community, its sleek façade concealing a nest of ruthless profiteers and worshippers of Mammon. But there was another Bristol: home to the largest dissenting community in Britain, with sizeable communities of Baptists, Congregationalists and Presbyterians and the nation's largest urban community of Quakers. These were among the wealthiest traders in the city; but they were also passionate campaigners for the abolition of slavery, happy to trade for no profit in the East Indies sugar that undercut the slave trade and brought ethical products to communities such as Beddoes' Shropshire. They were firmly opposed to war on moral and religious grounds, but they had become careful over time not to draw attention to their beliefs, demand their rights too loudly, or challenge patriotic calls to Church and King.

The outbreak of war had exposed the differences between these two Bristols, and at the same time highlighted the slow decline in the city's fortunes that now threatened to become a crisis. Though the city still presented a brash and bullish face to the world, the truth was that its importance to Britain's trade had peaked a generation earlier, and that the great prizes had been ebbing away for decades to new boom ports such as Glasgow and Liverpool. The Bristol docks ran right into the centre of the city, a very convenient arrangement; the handsome new urban centre, radiating out from the grand terraces of Queen's Square, had brought spacious, modern town houses to the backs of the wharves and warehouses; sledges laden with goods trundled up and down its business district of narrow streets and hills distinctively studded with steeples. But the Atlantic trade was now arriving in larger and larger ships, which could no longer reach the berths in the heart of Bristol: forced to dock several miles away at the mouth of the Avon, many were choosing other destinations. Plans for a floating dock to accommodate these vessels had been under discussion for years, but the Bristol Corporation, which controlled trade development, was paralysed by competing bids and vested interests, and by the complacent voices that continued to insist that large cargo ships were a minor trend that would prove insignificant in the long run.

* * *

Beddoes' search for laboratory premises, and for a new home, focused on the small spa resort of Hotwells, huddled in the steep and narrow defile of the river Avon. At this point, before it was dynamited to make it more accessible, the Avon Gorge was a startling wonder of nature. Steep cliffs exposed dramatic strata of carboniferous limestone, with mossy crags draped in trailing creepers and honeysuckle, and sheep grazing on handkerchief-sized patches of lush grass seemingly suspended in mid-air among the almost tropical greenery. Invalids who had toured Europe for their health frequently declared that they had traversed the Alps without ever seeing such a sublime spectacle. Adding to the sense of natural wonder were the hot springs that gushed into the Avon, exposed at low tide and picturesque in winter as wisps of steam wreathed around them. The springs had been held to have curative powers since medieval times, and since 1696 they had been enclosed in a brick bathhouse fed by a pumping system. During the eighteenth century a visit to the baths had become a popular excursion, and the narrow bank at the foot of the cliff was now a river promenade shaded with lime trees, behind which hotels and residential care homes fought for space with souvenir shops selling bottles of the spring water.

Hotwells, like Bristol proper, was in a slow decline after its peak a generation before. It had built its prosperity as a satellite to the vastly popular spa town of Bath, from where visitors could cruise a few scenic miles down the Avon to appreciate its pristine beauty, bypassing the wharves, warehouses and factories of downtown Bristol. Once disembarked, they could wander up and down the promenade or take cruises round the harbour, have strawberries for tea and watch firework displays on summer evenings. But by the 1790s its genteel attractions were obliged to compete with the desperate spectacle of its invalid residents. The hot springs and boarding houses had become a last-chance saloon, the grim terminus for those for whom all other treatments had failed and who could not afford to travel further: 'a Hotwells case' had become physicians' argot for a patient whose condition was hopeless. The promenade was crowded with cadaverous patients in the final stages of consumption, and the Strangers' Burial Ground up the hill in Clifton was crammed with the bodies of those for whom the last miracle cure had failed. The proprietors of the hotels and guesthouses often doubled as funeral directors, sometimes burying deceased guests so promptly that, by the time their families arrived, there was nothing left of their relatives but the bill for bed, board and last rites.

Hotwells teemed with local doctors and itinerant quacks who touted the hot springs as a cure for gout and kidney stones, scabies and diabetes; Beddoes regarded the waters as worthless except as a magnet for a clientele who could sustain both his medical practice and his need for experimental subjects once his project took shape. From his rented rooms in Clifton, he wheezed up and down the precipitous paths to the river front, where new terraces and squares had begun to spring up, backing onto the cliffs behind the row of hotels in front of the quay. Before long he had found an ideal property: a tall, large-windowed, modern town house in a new development a short, steep climb up the cliff, with an open view to the harbour and facing into its breezes. Its address was Hope

Square, which sounded an optimistic introduction to invalid visitors, though it was in fact named after Lady Henrietta Hope, who had funded the building of a chapel on one of its corners.

Hope Square's residents were not happy at the prospect of being joined by Dr Beddoes. A pneumatic laboratory sounded like a troublesome neighbour, and they feared being poisoned by invisible but toxic fumes or having their homes damaged by fires or explosions. Like all Hotwells residents, they were alert to the risks of their backwater becoming a haven for patients with consumption and other contagious and fatal diseases. Beddoes' political reputation had also preceded him, and they had no wish, particularly in time of war, to welcome into their community a notorious enemy sympathiser. Local objections were mounted, and delays followed; Beddoes, without friends in the city, cast around for a respectable local figure to support him. He did not have to look far. James Keir replied to his request, enclosing a letter of introduction to Sir Richard Edgeworth, who was currently resident in Clifton as his young son, Lovell, was being treated for consumption there.

Edgeworth had known Keir, and had been a central figure in the Lunar circle, since 1766, when he had made an unforgettable first impression on Erasmus Darwin with his combination of scientific tricks and conjuring ('the greatest conjuror I ever saw! He has the principles of nature in his palm, and moulds them as he pleases. Can take away polarity and give it to the needle by rubbing it thrice on the palm of his hand . . . astonishing! diabolical!!!').[4] The two had met through their shared fascination with a mechanical model of the solar system called the Microcosm that was on display in Chester, and they had entered into a vigorous correspondence about carriage design. Beyond their many shared passions, however, Edgeworth's social world was very different from that of his new acquaintances: he was an aristocrat, and Irish, though in most respects typical of neither group. The son and heir of a gentleman with an estate at Edgeworthtown in County Longford, he had been sent to Trinity College Dublin, where he had embarked on the career of a dissolute young buck with such furious determination that his terrified father had removed him to Oxford. There, he had eloped to Scotland with his landlady's daughter, become a father and studied law intermittently, though he was always far more preoccupied with designing and building a procession of mechanical contraptions, ranging from carriages and orreries to a prototype telegraph that operated by semaphore.

Falling in with the Lunar circle focused Edgeworth's innate mechanical genius. By the 1770s he was working on James Watt's steam engine designs, standardising the measurement of horsepower and taking out patents on a 'portable railway or artificial road' that could be rolled out by a track-laying vehicle.[5] He also became, after the death of his father in 1769, the lord of a ramshackle and politically restless Irish estate, which he spent much of his time reforming and restoring after years of neglect. In the 1780s, when Volunteer militias began to foment calls for Irish independence, he took his tenants' grievances to Parliament and supported the campaign for Catholic emancipation. He also, as time went on, became consumed by his constantly expanding family. By the time he arrived in Clifton in 1793, he was married to his third wife, and his brood of half-siblings extended from his oldest son, Richard, only four years younger than Beddoes, to the four-year-old Lovell. (There was another wife yet to come, who would bear him six more children.)

Edgeworth's succession of wives was not the result of inconstancy but of tragic illness and, in particular, consumption. His second wife, Honora, had contracted the disease in 1778; the family had moved to Beddoes' home town of Shifnal so that she could be attended by Erasmus Darwin, but she lost her struggle for life two years later. Her daughter, also Honora, was consumptive too, and died in 1790; the move to Clifton was, as for so many, a final attempt to prevent young Lovell from following his sister. Uprooting the family from their Irish seat brought with it many stresses, but Clifton was an agreeable second home. When Edgeworth tired of the scenery and needed stimulation, he could visit Bristol's well-stocked library or work on schemes to improve the Derby cotton mills; his third wife, Elizabeth, sister of the late Honora, could take a ferry over the Avon and strike out on solitary and bracing walks across the downs. They had, as Edgeworth often contrived to do, rescued a happy life from an unhappy situation.

For Beddoes, alone in a new town, this genial and bustling tribe were absorbing companions. He and Edgeworth shared their mutual friends and interests, and rapidly discovered more. Both were passionate about education and the need for its reform: Edgeworth had been a close friend of Thomas Day, and indeed *Sandford and Merton*, a touchstone for Beddoes' theories, had been written in response to Edgeworth's complaint that he could find nothing suitable for his children to read. Edgeworth's aristocratic credentials smoothly resolved the disputes around Hope Square, and Beddoes moved in, Joseph Sadler coming down from Oxford to help him set up his laboratory. As spring progressed into summer, Beddoes became embedded in the family, a favourite with the Edgeworth children, who 'jump about me, entreat me to go to Ireland, and consider my occasional absence from dinner a serious calamity'.[6] He sought out a range of practical toys for them to play with: small gardening tools, printing blocks and looms, pulleys and carts, with which they could take their first steps in experiment and applied science. 'Most playthings', he pronounced, are 'worse than useless . . . the flush of delight, arising from the first impression, cannot but be transitory; and no sooner does the little possessor examine into the structure of his new acquisition, than he flings it aside, or tramples it with his foot, as if to revenge himself upon it for belying the promise of its exterior'.[7] Play bloomed into a new project: a factory to mass-produce 'rational toys', as he called them, that would harness young heads, hands and natural curiosity to the project of deciphering the world.

*   *   *

His early days in Bristol also brought Beddoes into the company of Tom Wedgwood, the youngest son of Josiah and, with his older brothers John and Josiah junior (or Jos), heir to the family's vast pottery fortune. Tom was currently spending most of his time at Cote House, a seventeenth-century estate on the downs beyond Bristol that John had recently bought as a home for his young family. Precisely who made the connection between Beddoes and Tom Wedgwood is hard to establish because there are so many plausible candidates: Beddoes was already acquainted with the Wedgwoods through his Shropshire circle; Erasmus Darwin had been Tom's doctor when he was a child; James Keir, who had introduced Beddoes to Edgeworth, was equally well acquainted with the Wedgwoods; Richard Edgeworth himself had corresponded with Tom in 1790, asking for technical help with a scheme he was developing to improve the ventilation in Irish prisons by passing heated airs through them.[8] Given the close weave of the Lunar tapestry, and the coincidence of their individual passions, it would have been remarkable if the pair had not met as swiftly as they did.

Tom, at twenty-two, was a strikingly gifted young man, but also in a moment of profound crisis, and Beddoes' arrival would do much to shape the next phase of his life. He had been a child prodigy who, hothoused by a solitary home education, had reached an extraordinary proficiency in mathematics before going on to Edinburgh University at the age of fifteen, where Newton's *Optics* awoke in him an all-consuming appetite for science. He returned to the family home of Etruria in 1789, settled into one of the best laboratories in the country, and began a life of almost uninterrupted experimentation, some of it related to pottery manufacture but much of it developing his theories on the nature of heat and light. In 1792 he had two papers read before the Royal Society in London; the second, on the temperature at which different bodies become red-hot, came close to establishing the general theory of incandescence that would not emerge fully for another fifty years.

But Tom's qualities included not only a radiant intelligence but an acute and mysterious propensity to illness, and the harder he worked, the more he was tormented by blinding headaches and crippling stomach pains that seemed to be building towards a general constitutional and nervous breakdown. He struggled on, determined to conquer both chemistry and pain, and driving himself to ever higher pitches of both: at the point when he finally collapsed, he had spent six months trying to find a way to suspend a thermometer in a vacuum so as to be able to measure accurately the heat received from the rays of the moon. A procession of doctors bled and purged him, but doubts began to cloud the expectations of his future role in the family business. His older brothers were cheerful and dependable, but lacked Tom's evident genius, and Josiah had for many years implicitly looked to his youngest and most fragile son to provide the spark that would maintain the Wedgwood name in the first rank of industry.

In an effort to calm the stresses that were clearly consuming him, the family suggested that Tom should travel to the Continent for his health, and he joined another Lunar son, James Watt junior, for a trip to Paris. Tom was an openhearted and optimistic supporter of the French Revolution, but young James revealed himself a 'furious democrat',[9] echoing the mob's demands for the blood of the king and Lafayette and swearing to sign up with the republican army if the aristocracy threatened the revolution's progress. The pair tramped around the filthy streets of Paris, shunning the company of British expatriates, speaking next to no French, losing themselves in the crowds of sans-culottes, joining the ominously militarised Bastille Day celebrations in the Champ de Mars, and glowering at the Tuileries Palace, 'shut up by the King's order, though they are a national property'.[10] These were scenes that made a lasting impression on Tom, but the adventure failed to arrest his physical decline, and by early 1793 it was clear even to his father that his condition was too serious for him to take up the responsibilities envisaged for him in the family firm. In April father and son dissolved the business partnership into which they had solemnly entered on Tom's sixteenth birthday and he retired to Cote House where, bedridden and shattered, he first made Beddoes' acquaintance.

The crisis in Tom's health meant, as it had done periodically since his childhood, that his life had shrunk down to a narrow and largely internal world, where the only sensation was pain. Headaches would keep him pinned in darkened rooms for days; even in his better moments he would often be confined to a bath chair, wheeled outdoors into the shade where his greatest pleasure was to watch children playing in the sunlight. The fear that this might be his condition for the rest of his life made him prone to long depressions and fits of acute despair. These were made even more painful for those around him by the extraordinary sweetness of his nature, his courage and disdain for self-pity, and his irrepressible instinct to turn his own suffering to the benefit of the rest of the world. Deprived of his chemical laboratory, he took his illness as a cue to turn to the inner laboratory of his own mind, a direction in which Beddoes' project would eventually follow.

Tom had long been in the habit of filling notebooks with observations which, under the influence of David Hartley's theory of association, increasingly coalesced round the nature of pleasure and pain, and the ways in which they are created in our minds by the endless tiny iterations of stimulus and habit. 'Might not all painful feeling', he wondered, 'be repressed while the opposite is cultivated?'[11] Many of our pleasures are first perceived as unpleasant until a taste for them is acquired: the acidity of wine, for example, becomes pleasant only once we appreciate the broader complex of sensations of which it forms a part. By taking conscious control of this process, might we not learn to find pain more bearable, even joyful, as the martyrs claimed? 'Disagreeable events', he observed, 'affect you ten times more in the anticipation than they would in their realities',[12] and he compiled long lists of techniques for distracting the mind from its tendency to dwell on suffering. Like Beddoes and Edgeworth, he sought to turn these insights into schemes of education that might engineer a child's nervous system to recognise and dwell on pleasure more readily than pain. With his nephews and nieces, and in his notebooks in the dark hours, he experimented with techniques for creating beings of happiness.

But, turned in on itself, his keen observation also became a powerful self-lacerating tool, and his reflections were merciless in their self-criticism. 'Men of study, beware of solitude', he warned. 'It invariably begets conceit.'[13] Arrogance, pomposity and pride were all enemies of true self-knowledge; humility, diffidence and respect for others were the essential preconditions of true happiness. 'Never was there yet a human being who had the honesty to declare all the sources of his pleasures', he insisted.[14] For example: 'What is it we wish in exciting the sympathy of others with our sufferings?' 'That others should be almost as unhappy as ourselves' was his verdict.[15] The ruthlessness with which he sought to gouge insights out of his own pain gave him something of the quality of an Enlightenment saint, an exemplar for those around him who hoped to change the world for the better.

Tom's relationship with Beddoes rapidly evolved into a cat's cradle of shared passions, collaborations and mutual dependencies. Tom became Beddoes' patient, and in his worst periods of illness his sickbed would be set up in

Beddoes' front room. The younger man's condition was a mystery that came to obsess them both. Since childhood, Tom had had countless doctors who had treated him according to wildly divergent diagnoses: some for headaches and nervous conditions, others for a gastric syndrome, still others for hypochondria or a nebulous condition of the nerves in some way generated by his obsession with work, or by the stress of succeeding his father at the helm of a vast business empire. If Beddoes was to be more than an empiric, shuffling remedies by trial and error, he would need to crack the code of Tom's inscrutable torments.

And just as Beddoes naturally became Tom's personal doctor, Tom naturally became, over time, the most dependable financial supporter of Beddoes' project. He would be ambassador and talisman of the mission to transform medicine through chemistry, which would inevitably merge with the struggle to rescue him from the condition that was so pitilessly consuming him. The two men would be doctor and patient, beneficiary and patron, professor and protégé, surrogate father and son; Tom would be the first but by no means the last of the brilliant representatives of the younger generation who would form a circle around Beddoes in Bristol, and take his work in directions that he had thus far not even begun to imagine.

* * *

Bristol had never seen a doctor quite like Thomas Beddoes. His Edinburgh degree and subsequent Oxford tenure, combined with the patronage of Erasmus Darwin, marked him as a significant arrival on the provincial scene, and his private practice extended rapidly through and beyond the Edgeworth and Wedgwood circles into the wealthy clientele to which most of Bristol's physicians aspired. But in Beddoes' case this lucrative trade remained only a means to developing his project for the benefit of the vast majority of Bristol's population: the poor. This far larger and less profitable constituency currently had two options for medical care. One was the 'vulgar' or 'vernacular' medicine of the non-specialist: folk remedies culled from herbals, the empiric compounds available from quacks and itinerant pedlars, bloodletters and inoculators, family traditions, old wives' tales and self-medication. The other was a hospital, the Bristol Royal Infirmary, which had been founded in 1737, underwritten by charitable donations and the combined poor-relief funds of the city's seventeen parishes. Here, the poor could receive free treatment from the surgeons and apothecaries' assistants who had, over the years, turned it into an unofficial teaching hospital.

Beddoes disapproved heartily of both of these options. Vernacular medicine, driven as it was by a money-grubbing and unregulated market, was, in his view, simply an open invitation for the unscrupulous to prey on the vulnerable. The apothecary's range of cordials, lozenges, balsams and plasters was advertised against everything from fevers to rickets, kidney stones to scurvy, but most of what was on offer was worthless even as a palliative, let alone a cure. The herbals, such as Nicholas Culpeper's, consisted largely of common plants – borage, rue, garlic, wormwood – whose administration might soothe the patient, or perhaps the carer, but had no true power in a medical crisis. Mixed in with these was arrant superstition: Culpeper still maintained, for example, that a cake of the sick person's urine mixed with flour and fed to a dog would pass the illness over to the animal.[16] Most people had learned to get by on haphazard and idiosyncratic self-medication regimes that were, in Beddoes' long experience, morasses of crackpot theory and self-delusion.

He was equally sceptical of the Royal Infirmary, largely for the reasons he had expressed in his pamphlet in Cornwall two years previously: it was an arrangement for the benefit of the doctors, not the patients. His scepticism was shared by the Bristol poor themselves, who hated the hospital and used it as little as possible. There was no such thing as a hospital for rich people, who were treated in their own beds. To be taken in to the Infirmary was to be removed from the support of family and friends and surrounded by strangers who discussed your condition in a foreign language, and who were clearly looking for subjects on whom they could either experiment or impose their fixed ideas. One doctor in the Infirmary in the 1770s, Dr John Paul, was well known to ask every patient whether he was a Bristol man; if he replied that he was, Paul would assume that he drank ale and smoked tobacco every night, and immediately order twenty ounces of blood to be let: 'the first thing to do is to let some of that run out, and then we shall see what else is the matter'.[17] Nor was it lost on the poor that doctors were dependent on cadavers for anatomy practice, and that once in the Infirmary they were in the care of people to whom they were more interesting and valuable as corpses, and who had no qualms about breaking the age-old taboos about the sanctity of the dead.

Beddoes, then, arrived in Bristol trailing more threat than promise for the medical scene. If his plans for pneumatic therapy prospered, he would be a plague on both these houses, draining customers equally from the quacks and from the established physicians. But he would be obliged to form common cause, on some levels, with his colleagues among the physicians, of whom he would soon be easily the most famous. He would be generous to those who wished to collaborate with his researches, and always ready to encourage, talk up and publish the experiments of others; but in doing so, he would be riding roughshod over long-established codes of professional conduct. Before his arrival, Bristol's medical establishment had been a close-knit and discreet association, publishing nothing beyond the odd learned medical tract, keeping well away from politics and quietly collaborating to shore up their professional network. Beddoes was certain to make enemies; it was yet to be seen whether he would also make friends.

Meanwhile, work in Hope Square was progressing slowly but steadily. By the end of May, he and Sadler had set up a laboratory that could produce factitious airs, and he had begun to test them on kittens. But there were more refinements to be made to allow accurate measurement of the dosage and purity of the airs, and to determine the best design for a room that would need to accommodate both laboratory and patient. And once these technical issues were resolved, there were wider questions about the protocols for working with human subjects. The very act of experimenting implied the possibility of error: if Beddoes' intended subjects were frightened away from the Infirmary by the prospect of postmortem dissection, why should they trust his untested pneumatic therapies? Should subjects be paid – and if so, would this attract the genuinely sick? And how acute should their condition be? If they were not seriously ill, there would be little opportunity for the remedies to work; and if they were, might the experimenter not seem to be preying on the desperation of the sick, and risking their death if the experiment failed?

At the time of his arrival in Bristol, Beddoes was seeking to strengthen his position with two new publications, one aimed at the poor and the other at the medical profession. The first, A Guide to Self-Preservation and Parental Affection, was in a small chapbook format, cheaply priced for the popular market and announced as a new work from the author of Isaac Jenkins. It was a simple self-help guide, intended to disseminate medically sound advice to the poorer classes, and its focus was not on the treatment of diseases but on their prevention. It broke the physicians' professional code by demystifying the art, confiding to the reader that most doctors' diagnoses are 'merely in the way of trade', and only designed 'to give a high notion of his skilfulness'.[18] It stressed the importance of eating healthily, dressing properly and avoiding sudden transitions from heat to cold; it warned against the adverse effects of spirituous liquors and proclaimed the cost-free benefits of cleanliness and fresh air, arguing that those 'neglecting to open their windows, are drawing in a foul, tainted air, a great part of their time, by which means some disorders are brought on, and others rendered worse than they would naturally be'.[19] He included as evidence for this the case of a gentleman who, when stricken by a violent attack of influenza, slept with his bedclothes off and awoke cured, with 'no more fever or disagreeable feelings', an experiment Beddoes had in fact tried the previous year on the only subject whom he could ethically commandeer: himself.[20] Here was a doctor the poor could trust: one who was not attempting to fleece them or corral them into the hospital, but sharing his experience and learning with them to help them avoid the necessity.

By the spring of 1793 *Observations on the Nature and Cure of Calculus* had already sold out several print runs: a success, Beddoes claimed to Giddy, 'almost unexampled among medical books in this country'.[21] Now that he had the ear of the profession, he moved to expand support for his new project. His new bulletin emerged under the title *A Letter to Erasmus Darwin MD on a New Method of Treating Pulmonary Consumption*, making full use of Darwin's imprimatur and publishing his warm letter of the previous year. In private Darwin was rather more circumspect, regarding Beddoes' project as 'very ingenious, but experiment alone can show whether it will succeed'.[22] This was an objection that Beddoes was eager to address, and Darwin's letter was already buttressed with a seventy-page update on his progress. He was, he announced, on the brink of a major breakthrough in the treatment of consumption, though equally 'I cannot but be aware that it would be extremely dangerous and impolitic to speak of it with greater confidence than it deserves'.[23]

By proceeding so publicly with such provisional results, Beddoes was well aware that he was setting himself up to be criticised not merely for his medical theories, but also for his methods. 'I must expect to be decried by some as a silly projector', he acknowledges, 'and by others as a rapacious empiric'. He cannot but be seen as a projector: since his theory of pneumatic therapy, if proven, would provide a defining boost for his personal enterprise, he must expect suspicion that his experiments are merely part of its publicity campaign. What precisely is the difference between his case histories and supporting testimonies, and the anecdotes of miracle cures plastered on the bottles of quack remedies? Giddy, with the sober objectivity on which Beddoes was coming to rely, had registered his concerns on precisely this point; Beddoes had replied that 'your idea of my plan appearing empirical occurred to myself, and of course I took much pains to guard against it'.[24] Beddoes' pains begin, here as elsewhere, with a quotation from Francis Bacon on the title-page: he is not trying to dazzle with mystery or to bully with spurious authority, but is working within an open framework of sceptical experiment applied transparently to testable hypotheses. If his theories prove false, he will be the first to know; and even if he wished to conceal the fact from his public, he would not be able to.

But although what he is presenting is a tentative work-in-progress, he already has two dramatic discoveries to announce. The first is a radical new classification of consumption, Brunonian in its sweep, that rejects the 'great multitude of species and varieties' of the condition in the medical literature as 'founded upon states of the body merely imaginary'.[25] The condition may manifest itself in a huge diversity of forms; it may attack the blood, the bones, the skin, even the brain; but at its root are always the same physical signature and cause, the lesions on the tubercules in the lungs. And his second discovery is that this root cause can be treated in a manner hitherto untried. 'Several years ago', he informs the reader, 'a firm persuasion settled upon my mind that the system might be as powerfully affected by means of the lungs as of the stomach. And the more knowledge we have acquired of elastic fluids, the more my opinion has been strengthened.' With the rediscovered researches of John Mayow, and Edward Goodwyn's meticulous investigation of the respiratory process, the groundwork has been laid for an entirely new species of gaseous medicine. Pneumatic chemistry has given us 'the command of the elements which compose animal substances'; now, 'it is the business of pneumatic medicine to apply them with caution and intelligence to the restoration and preservation of health'.[26]

This original therapy may be in its earliest stages, but there are already positive results to report. Beddoes details the treatment of Mr Crump's son in Shropshire the previous winter, and his own self-experiments with oxygen, and proceeds to set out the research agenda that he believes to follow logically from the facts thus far known. Hydrogen, he proposes, will be invaluable 'to reduce air to a lower standard' for the treatment of hyperoxygenation in consumption and other diseases; he has already established that 'animals may be gradually iniured to air with less oxygen than usual'.[27] Oxygen, by contrast, shows promise in offering relief from 'a considerable variety of diseases' that arise in whole or part from its suppression, including perhaps typhus, hysteria, diabetes, liver damage, palsy and ulcers. Once these therapies are proven by experiment, the strange notion of doctoring by chemical airs will become commonplace, and 'a convenient small apparatus for procuring and containing oxygen air' will, he predicts, 'soon come to be ranked among the ordinary items of household furniture'.[28]

*　*　*

Beddoes was by now travelling a great deal as part of his medical practice – 'traversing England & Wales in order to secure a subsistence at least', he grumbled to Giddy[29] – but he continued to spend as much time as possible in Clifton with the Edgeworths, where the now-familiar social pleasures had been joined by an enticing new one. It was during the time he spent nursing young Lovell that he first began to pay attention to Anna, Edgeworth's twenty-year-old daughter from his first marriage, whose 'conduct to her sick brother was so affectionate & so unaffected as to engage my attention extremely'.[30] Beddoes made formal and evidently rather stilted conversation, choosing 'those uninteresting topics, which are so well calculated to relieve the distress of persons who must say something to each other and know not what to say'; but Anna had a way of turning formality to intimacy, and Beddoes found himself having far more interesting conversations than he had intended. He discovered that Anna's 'thirst for knowledge was very ardent', and that 'her opinions on politics and religion coincided with my own'; furthermore, that she had an 'extraordinary proficiency in arithmetic', a fine writing style and 'morals . . . equal to her understanding'.[31] By the end of May it was clear to both of them, and probably to everyone around them, that their companionable conversations were developing into a serious courtship.

Anna was an unconventional child of an unconventional family, bright and forward in conversation and able to draw Beddoes out of himself in ways that no woman had done before. She was witty and high-spirited, fond of poetry and theatre and music, and a natural companion to the young half-siblings to whom she was a combination of mother and sister. But her upbringing had in many ways been a troubled one, quite different from the carefree liberality that her father was now cultivating with his third brood. Her mother had died during her infancy, and Honora, Edgeworth's second wife, had treated Anna with strict discipline, forcing her at the age of three to dress herself and make her own bed, and withholding her breakfast if she failed. The succession of stepmothers, family deaths and contradictory educational regimes had left Anna less sure of herself than she seemed, and with more complex needs than the easy society of her family tended to acknowledge.

Although she and Beddoes were obviously attracted to one another, their liaison developed with a rapidity that suggests each had been thinking seriously about marriage before they met. Anna, away from the rural isolation of Edgeworthtown for the first time, was eager to find a way of making the bustling cultural life of Bristol her permanent one; Beddoes, for his part, had given intimations that he was ready for a romantic attachment before Anna came along. In many respects it was a rare match between two idiosyncratic characters whose unconventional expectations of marriage limited severely their choice of partner. Both held the view that a wife should be not subservient but equal, and that money was immaterial to the match; but both also had their doubts about the union, and lacked the experience to know whether these were really doubts about the institution of marriage, or about each other, or about themselves. Behind Beddoes' sociable and eloquent exterior, he was still a shy bachelor who had not visualised the practicalities of married life in any great detail, and who worried 'if it is not a law to which all married men are doomed, that in time they shall grow tired of a woman'.[32] Equally, the gaiety of Anna's manner concealed a more anxious and highly strung side: her light-hearted teasing expressed in part a need for attention that could become waspish and neurotic when not indulged, and she may have been anxious from the beginning that, once married, she would disappear into the background of Beddoes' busy life. In each case, their delight at finding an honest and affectionate match was shaded by apprehensions about the future that it entailed.

The fraught negotiations that were bound to follow from a tanner's son claiming the love of a gentlewoman and heiress were handled by Richard Edgeworth with his customary combination of charm and frankness. He broached the subject after dinner one evening in May 1793 by remarking that he had always wanted his daughters to choose their own husbands, and that their social status was irrelevant: 'if the lady chooses to match with the cobbler, I would do all in my power to procure him the best stall in the street'. Beddoes felt that he had been treated 'in the kindest and most liberal manner',[33] and welcomed with an open heart into a family who had become dearer to him than his own. In private Edgeworth was indeed affectionate; but he was also sceptical and prescient. 'The object of Anna's affections', he wrote to one of his more conservative friends, 'is a little fat democrat of considerable abilities, of great name in the scientific world as a naturalist and chemist – good humoured good natured – a man of honour and virtue, enthusiastic and sanguine and very fond of Anna'. As to his prospects, 'if he will put off his political projects till he has accomplished his medical establishment he will succeed and make a fortune – but if he bloweth the trumpet of sedition the aristocracy will rather go to hell with Satan than with any democratic Devil'.[34]

However relaxed Edgeworth was about Beddoes' social origins, discussion of money was unavoidable, and he presented his calculations to his prospective son-in-law promptly and openly. He suggested that 'if we would raise £500 a year between us we need not hesitate a moment on this account', but that Beddoes' medical practice was, at this stage, too unpredictable a source of income to include in the projections. Calculated at roughly double the interest on her eventual fortune, Anna was worth around £100 a year; Beddoes, expecting a future inheritance estimated at between £3,000 and £6,000, felt he could guarantee another £300;[35] but the remaining £100 was a hostage to speculations that Edgeworth had gently ruled out of the equation. Beddoes was quick to appreciate overall 'how much below the aristocratical estimate this is', and not for the last time, had to turn to Giddy to make up the shortfall. His friend was happy to assist him in buying his bride, a gesture that subsequent events would render rich in irony.

It had been a whirlwind courtship, and settled with the briskest and most amiable efficiency imaginable; but no sooner was it sealed than Beddoes and Anna were whisked apart. Reports of political unrest in Ireland had become alarming. Since the declaration of war with France, underground sectarian activity had intensified: Catholic societies were, by various accounts, forming militias, drilling covertly and stashing away pikes to support a rumoured French invasion, and British forces had begun brutally dragooning the countryside, arresting and torturing suspected republican conspirators. Edgeworthtown was undefended, occupied by nobody but Edgeworth's tenant farmers; like Giddy in Cornwall, it was his job to call up and deploy local militias if needed. He could no longer afford to arrange his affairs around young Lovell's health.

The Edgeworths' departure, according to Anna's sister Maria, 'seemed particularly to grieve and alarm Dr. Beddoes';[36] but Anna, perhaps to Beddoes' surprise, was sanguine about a temporary separation, telling him that 'she thought it prudent for people who liked each other to separate for a time in order to try whether the sentiment is durable'.[37] She could use the time, she told him, to improve her housekeeping skills by taking over the estate's management from her father. If Beddoes felt that her suggestion concealed any uncertainty about his proposal, he concealed it in turn from his friends, to whom he presented her disappearance as eminently sensible: 'good reasons', he told Giddy, 'make her inflexible upon this point'. Yet he felt confident enough to break the news of the engagement to his parents, assuring them that 'Anna E. is just & generous & as free from greediness as it is possible for man or woman to be', and that 'as soon as I can afford it, I will cease to take anything or at most about £100 a year from you'.[38]

*   *   *

As the summer of 1793 turned to autumn, it was not only in Ireland that the political temperature was rising. France was a turmoil of rumours and conspiracies, plots and purges: since the assassination of Marat in July, the Committee of Safety had come under pressure from all sides to take ruthless measures against the provincial rebels, covert monarchists and fifth columnists of foreign powers whose hand was detected in the growing public disturbances. After a long dry summer, bread supplies were beginning to run short; demonstrators disrupted the constitutional deliberations of the national Convention, brandishing newspapers in which Jacobins, Dantonists, Hébertists and militant *Enragés* claimed daily that their faction was the last bastion of patriotism against the lies and duplicity poisoning the nation. It was the scenario that Beddoes had dreaded: he had long distrusted the 'shuffling, smooth-tongued villainy of Robespierre, the dark, sanguinary ambition of Marat', and feared that the consequences of these intrigues would be a revolution without legitimacy, 'upheld merely by a handful of factious men' doomed to be swiftly and brutally replaced by another.[39] The spirit of innovation had been released, and iniquitous authority had been liquidated; but, just as its enemies had predicted, the great experiment was spiralling into anarchy and terror.

In Britain it was becoming ever clearer that those who promoted political innovation and dissent were to be met with the full force of the law. In Scotland, where opposition to the costly and unpopular war had been building since the spring, the home secretary, Henry Dundas, had spent the summer mobilising a network of informants to track down the authors and publishers of pamphlets considered seditious, and was now mounting a series of prosecutions unparalleled in their severity. In August, Thomas Muir, a young lawyer who had organised the Edinburgh branch of the reform-minded Friends of the People, was convicted on flimsy evidence of sponsoring an *Address of the United Irishmen* calling for self-government; in September, Thomas Ffyshe Palmer, a wealthy Eton- and Cambridge-educated minister who had been converted to Unitarianism by the writings of Joseph Priestley, was accused of authoring an anonymous pamphlet that proposed petitioning Parliament against the war. The juries who found the men guilty were astonished at the sentences handed down by the court: instead of the customary few days in jail, seven and fourteen years' transportation respectively. For both men, it was close to a death sentence: Muir would eventually escape from Botany Bay, travelling via Havana to France and losing an eye en route, while Palmer would die a Spanish prisoner of war on the shores of Guam in the South Pacific. A string of further convictions served notice that, however well intentioned their aims or genteel their proponents, dissenting opinions were henceforth to be treated as sedition and adherence to the enemy in time of war.

In Bristol, as the chill of autumn took hold, there was a new and ugly mood on the streets, though it had little overt connection to war or treason. Its flashpoint was the main bridge over the Avon, which had recently been restored and whose trustees had been authorised by an Act of Parliament to raise tolls on it until they had recouped their expenditure, around £2,000. But as the date for free transit approached, the trustees announced that costs had not yet been met, and that tolls would continue to be raised for another year. Complaints were ignored, the trustees refused to submit their accounts for inspection and the volume of dissent rose until, on the night of 19 September, a mob descended on the bridge and set fire to the tollgates. The following morning the trustees advertised rewards for information leading to the arsonists' arrests and new punishments for assaulting toll collectors, and began to construct new gates. On the night of Saturday, 28 September, a mob descended once more and destroyed the rebuilt gates; the Herefordshire militia was summoned, the Riot Act was read to no effect, and the soldiers fired into the crowd, wounding several and killing at least one as they fled.

Giddy had been forced to resort to a similar crude display of force to calm the Cornish mobs; but in this case it turned out to be the trigger for the Bristol Bridge Riots, as they would become known, to begin in earnest. The following day a much larger crowd appeared to demonstrate against the toll collectors, who were this time supported by city magistrates with the powers to read the Riot Act, call out militias and arraign lawbreakers. On the Monday the two sides squared off once more; the Riot Act was read three times, but the crowd refused to disperse. As dusk fell, the collectors abandoned the tollbooths, which were once more set alight. The militia, with the mayor behind them, advanced onto the bridge; the crowd stood their ground, and the troops opened fire. This time, the protesters were determined to hold the bridge, and the troops fired volley after volley. By the time the crowd had finally been broken up, at least ten had been shot dead and over fifty wounded.

Beddoes' newly adopted city was now faced with the most serious crisis in its history. The Bristol authorities had never been popular with the mass of its citizens: they were typically seen as an incompetent, corrupt and unaccountable elite who had the city's substantial if dwindling wealth parcelled up between them; but never before had they had the people's blood so visibly on their hands. The members of the Bristol Corporation, who ran the police and the court sessions, appointed the magistrates and had the power to call out militias, were largely the same individuals who composed the Society of Merchant Venturers, the administrators of the city's lifeblood, the shipping trade; and it was no surprise that the same individuals were well represented among the bridge's trustees. They were a conspicuous oligarchy, and an easy one to locate. On Tuesday, 1 October the citizens abandoned the bridge and massed in the centre of town. They descended on the Guildhall, owned by the Corporation and now serving as barracks for the out-of-town militia, smashed its windows, and moved on to the Corporation's headquarters at the Council House where they did the same. By midnight the Hereford militia, who were patrolling the city in large packs and with no great conviction, had decided to withdraw entirely. The following morning they were replaced by militias and dragoons from Monmouthshire and Wales; but this time the charred booths were not replaced, and no further attempt was made to collect tolls on Bristol Bridge.

As stunned Bristolians attempted to make sense of how their prosperous, contented and united city had exploded into carnage, whispers of ringleaders and Jacobin agitators naturally began to appear in official statements and press reports. But this line of explanation had a credibility problem: officially, Bristol had no such agitators. The previous year, when Davies Giddy had been forced to affirm the royal proclamation against sedition in Cornwall, the lord lieu-tenant of Somerset had been asked by Henry Dundas about conditions in Bristol, and had told him that 'the number of seditious persons, if there are any, is so small as to be by no means an object of the smallest apprehension or danger'.[40] The Corporation, having stonewalled these enquiries to avoid embarrassing government scrutiny, could not easily use nonexistent sedi-tionaries to square the circle between Bristol's harmonious self-image and the bloody scenes that had unfolded.

The story supported by the public feeling on the streets, and now expressed in a volley of briskly circulated pamphlets, was very different: it was one not of revolutionaries and ringleaders, but of a deep and long-standing rift between the city authorities and the citizens they governed. For the majority, the blame for the tragedy lay with the Corporation's greed, intransigence and panic. Eyewitness accounts featured no shadowy Jacobins, and only a very small number among the crowd who were responsible for the arson and violence: the majority were 'lookabouts and idlers', it being only 'natural that numbers of people should stop on one of the most populous thoroughfares in the kingdom'. Most of those killed, by the same accounts, had been shot in the back while trying to escape the violence. After the incident it had been the citizens them-selves who had taken the lead in restoring order: all had been 'extremely cautious not to afford the least pretext for numerous assemblies'.[41]

Yet there was more to justify the Corporation's mutterings of Jacobin infiltra-tion than it cared to admit in public. Its response to Dundas had been less than candid: there was at least one corresponding society in Bristol organising for political reform, and at least one bookseller carrying copies of Rights of Man in its cheap sixpenny edition. But the forces of Church and King loyalism were also massing in the shadows, and already indicating that they would be far more aggressive in taking control of the streets: the Corporation had received anonymous letters threatening that 'we are coming near 2000 harty strong rufins' and that 'all your dissernters houses shall share the same fate as them at Birmingham [sic]'.[42] Though they kept as silent about these threats as about their political reformers, Bristol's elite were well aware that they posed a far more serious threat to public order than any alleged nest of Jacobins.

The charred remains of the Bristol Bridge tollbooths were cleared away and the crowd resumed its peaceful bustle around the docks; but the mental and social scars of the violence were not so easily healed. Riots in eighteenth-century Bristol, although less common than elsewhere, had not been unknown; but these most recent ones had expressed a savagery that was recognised by both sides as something new, and perhaps a harbinger of worse to come. 'The present generation', Beddoes prophesied gloomily to Giddy, 'has much evil and very little good to look forward to'.[43] Perhaps he was beginning to sense that the next time the city erupted into conflict, he would no longer be watching from the sidelines.

* * *

In the meantime Beddoes' project was beginning to draw curiosity and support from the great and good. In December he was visited by Georgiana, duchess of Devonshire, who was residing at Clifton; she spent three hours in his laboratory in Hope Square, during which she 'minutely examined the whole apparatus'.[44] Georgiana had recently returned from a period of self-imposed exile on the Continent, during which she had given birth to a child of doubtful parentage; her legendary salon days were behind her, and she was attempting to reinvent herself as a patron of the sciences. She had a ravenous curiosity about the new geology, and a considerable collection of minerals and fossils that had attracted the interest of Joseph Banks, the president of the Royal Society, whom she had met in Naples where they were both studying the volcanic activity of Mount Vesuvius. On her return she had set up a chemistry laboratory in a back room of her London residence, Devonshire House, and she impressed Beddoes with her grasp of the subject. She in turn was impressed by his 'genius and good sense', though she added the qualifier, 'except in politics, in which he has neither judgement, taste or temper'.[45] Like Richard Edgeworth, she had a worldly sense of the hazards to which his trumpet of sedition was liable to expose him.

It was during Georgiana's visit that Beddoes was encouraged to expand his scheme to include not only a laboratory and surgery but a residential hospital, where patients might be able to respire mixtures of gases not only in brief treatment sessions but over long periods, perhaps even continuously. 'Six to twelve patients' under constant treatment, Beddoes decided, would advance his researches more rapidly than 'twelve years of private practice'[46] with outpatients submitting to occasional inhalations. This would, of course, increase the funds required for the project, but six or seven hundred pounds should still be adequate. 'Bad as the times are, could one not find benevolent people enough to assist in the execution of so grand a design?'[47] Georgiana, infected with his enthusiasm, departed eager to add her name to the project's supporters and to be kept informed of its progress.

Beddoes' technical ambitions were also expanding, and beginning to outrun his own and James Sadler's expertise. The laboratory at Hope Square was capable of administering airs, but its apparatus was improvised and cumbersome, producing them at varying pressures and purities and through laboratory pipes that were poorly designed for inhalation by sick patients. He decided to seek specialist help, and he aimed high. In March 1794 he wrote to the nation's foremost engineer, James Watt, asking him to design a customised pneumatic apparatus that could synthesise a range of gases and dispense them comfortably to patients, initially under supervision in his clinic, but eventually in their own homes.

Watt was not a hard man for Beddoes to reach. As with all the Lunar circle, the two men were connected by many mutual friends, and indeed Beddoes had by now become the main point of contact for the ageing and increasingly sedentary group whose habitual method of contact had been disrupted by the Priestley riots and their aftermath. Watt had collaborated for decades with Erasmus Darwin and James Keir; he communicated regularly with Joseph Black in Edinburgh, and Tom Wedgwood and his son James junior had remained intimate since their days in revolutionary Paris. But approaching Watt was still a delicate business: he had a reputation for prickliness in financial dealings and anxieties about patents and payments. The tone of Beddoes' request to him was carefully modulated, and in startling contrast to the confident claims that had begun to characterise his public dispatches. 'I do not believe in my own theories', Beddoes informed Watt candidly; 'for instance, I do not believe in the hyperoxygenation of the system in consumption'.[48] Such claims were necessary for the projector to catch the public imagination, and to give his research programme 'a plausible form' to test by experiment; but Beddoes was, at least in private with Watt, happy to confess himself a Baconian empiric, with no theory beyond that of following wherever experiment might lead.

Beddoes' timing was fortunate. Watt, now aged fifty-eight, was relaxing into semi-retirement with a career of spectacular success behind him and, finally, a secure cushion of wealth on which to rest. He had recently moved out of a rapidly expanding Birmingham, its growth predicated in no small part on his and Matthew Boulton's Soho works, into a large custom-built house among forty acres of parkland in outlying Handsworth. Both partners, as they aged, were reverting to type: Boulton still whirling around the city consumed in meetings and deals, while Watt retreated to the sprawling workshop in his new attic where he could lose himself, tinkering in solitude and surrounded by tools and instruments. It was turning into the happiest period of his life, with the exception of one deep and dark shadow: consumption. His fifteen-year-old daughter Janet, known as Jessie, had long been rendered an invalid by the disease, and her fits and discharges of blood were becoming more violent; her older brother Gregory, already showing brilliant promise as a chemist, was beginning to suffer the same symptoms. Beddoes' letter reached Watt at a moment when the cure for consumption was becoming a preoccupation, even an obsession, with the great engineer.

Beddoes' wait for a reply was interrupted by a happy adventure: a journey to Ireland to marry Anna at the Edgeworth family home. He found the tribe humming with energy as ever, and Edgeworthtown an oasis of good humour and stability in a country in turmoil. Most landowners lived in dread of uprisings among their tenants and infiltration by the Defender militias who were suspected of plotting with the French for an imminent invasion. Edgeworth was not immune to such concerns – as he put it to Darwin, 'an insurrection of such people, who have been much oppressed, must be infinitely more horrid than any thing that has happened in France'[49] – but he had for precisely this reason put his own estate on a relaxed and even democratic footing. He had swept away the predatory agents and middlemen who had been taxing the tenant farmers in his absence, and distributed their land according to need rather than by auction. He had abolished the tradition of compulsory and unpaid 'dutywork' at harvest time, made cash grants available for repairs and improvements, and collected rent a year in arrears rather than in advance. While other landowners were anxiously building high walls around their gloomy mansions, Edgeworth's modest and tidy three-storey house opened onto a large lawn on which his tenants' sheep grazed freely.

All this gave Edgeworth a great deal more work to do than most landowners, on top of which he was, as ever, working compulsively on a new invention: a 'Tellograph', as he christened it, which might be used to communicate news of rebellions or invasions. By the summer he would have sent messages across a twelve-mile test route to the town of Collon, and he was already devising a code far more sophisticated than the alphabetical system adopted by the early French experimenters, using pointers, numbers and geometric shapes to generate a vocabulary with seven thousand possible combinations. He would attempt to interest the Irish government in both the device and the signalling system for many years, without any success.

The wedding was an effusive occasion and, contrary to his habit, the local vicar even remembered to register it properly. Beddoes wrote to his parents from Ireland, his happiness clouded only by his anxiety that they might not be as warm and generous to Anna as her family had been to him. Nothing, he warned them, would 'hurt me so much as any sort of dislike or unkindness shown her by my relations'. He warned them in particular that Anna had no need to hear any disagreeable tales about their financial wrangles: 'I never told A.E. a word of the reasons I have to complain of many improvident steps of yours. I never shall for I love her too well to wound her mind with such recitals.' Anna had, he told them, 'always lived happily in a country where her father has been a kind of king';[50] even if his own family had been no such fairy tale, he prevailed upon them to refrain from breaking the spell.

By the end of April, Beddoes and Anna were back in Bristol as man and wife and setting up home in Rodney Place, a row of handsome Bath stone town houses set back from the main street that ran through Clifton village. Beddoes' surrogate father-figures welcomed the match: Darwin was delighted, and Edgeworth in no doubt that Anna was happily settled. The couple engaged a servant boy, whom Beddoes promptly taught to read and write. Anna took over the correspondence with Beddoes' parents, and her relaxed and cheerful tone replaced their son's prickliness. But, in a sign of things to come, their domestic harmony was interrupted almost immediately by an urgent medical call. James Watt replied to Beddoes' letter with a plea for help: Jessie's condition was slipping from bad to worse. Darwin had been attending her with increasing frequency, but they had exhausted all possible remedies. Might he come to Birmingham and attempt some pneumatic therapy?

The catalogue of treatments that had been tried on Jessie stands as an eloquent explanation for Watt's willingness to assist in Beddoes' scheme. She had been fed tinctures of foxglove, castor oil and bark tea, and sedated with laudanum; she had been exposed to shiver in the cold air several times a day all through the winter; she had been blistered and bled, and swung about on a rope to make her sick. By the time Beddoes arrived, she was limp and almost lifeless. For a week he stayed at her bedside, in the anguished ordeal he had relived countless times since his childhood vigil beside his grandfather. This time, however, he set up an alembic and drew 'fixed air', or carbon dioxide, into a bag for her to inhale, sometimes pure and sometimes mixed with atmospheric air. At times it seemed to benefit her and calm her hectic fever, at others she was too weak to inhale it; by the beginning of June she was dead.

Beddoes was shattered, profoundly concerned about Watt's mental state, and tormented by the anxiety that 'the inspiration of fixed air could be suspected of having done injury'.[51] But Watt, struggling as he was to cope with the enormity of his loss, was far from blaming the airs. In reply to Darwin's letter of consolation, he wrote: 'I have long found that when an evil is irreparable, the best consolation is to turn the mind to any other subject that can occupy it for the moment'.[52] He had seen the moments of relief that Jessie had seemed to feel from the pneumatic therapy, and his mind had begun to turn on ways of improving Beddoes' clumsy laboratory apparatus. He was not convinced by Beddoes' theories – a leap of faith that Beddoes had been at pains to stress was unnecessary – but, after he had lost Jessie, the experiment seemed to offer a sense of purpose and catharsis. 'I told you that I had turned my contemplations to the subject of medicinal airs', he continued to Darwin, 'not from any idea that I might understand the subject, but because nobody else does'.[53]

Bastille Day of 1794 turned out to be another momentous 14 July for Beddoes. He received a letter from Watt including not only drawings of his proposed range of custom pneumatic apparatus, but the news that it was to be manufactured by Boulton and Watt at Soho, the nation's leading manufactory – and that, breaking the habits of a lifetime, Watt was prepared to undertake production without applying for patents or making any claims on copyright.

Watt's drawings showed a portable stove, on top of which was mounted an alembic in which the the desired gas would be produced: oxygen by dripping acid on Exeter manganese, hydrogen by heating zinc in sulphuric acid, carbon dioxide by adding water to red-hot chalk. Above the alembic, a tube would carry the gas, through a neck that could be opened and closed, into a refrigeratory, a water-filled trough in which the gas would be cooled. From here, it would pass into a tinplate holder that would rise and fall with the pressure of the gas; the gas could then be drawn off into a portable bag, meaning that red-hot flasks and inflammable reactions no longer needed to be located directly at the bedside. The bags recommended were of oiled silk, soft but impermeable, treated with a dusting of charcoal to absorb chemical smells and impurities. The apparatus came in various sizes, with more complex parts such as double-lined and circulating refrigeratories for large-scale and difficult reactions, and accessories such as joint-sealing cement and a beehive-shaped cap that could be filled with gas for patients who were too weak to inhale. The standard model was priced at £8 15s 6d.

*   *   *

Watt's designs were precisely what Beddoes needed to tip his scheme from idea to reality. He had outlined his mission, shored up his theoretical base and given the public a taste of discoveries to come; now he was in a position to set the hare running with a prospectus and a business plan with which he could sign up subscribers and funders and begin his experiments in earnest. He began to activate every strand of the network he had assembled over the years in Shropshire, Oxford, London and Bristol: drawing up lists of potential subscribers, collating all his pneumatic researches to date, identifying the obstacles to progress and worrying away at modes of administration and therapy. 'I have met with great encouragement to prosecute this idea', he wrote to Tom Wedgwood, 'and even in these bad times, I entertain great hopes of raising a sum adequate to the purpose'.[54] He was now estimating his needs at between £3,000 and £5,000, but his subscriber base was growing to match them. The duchess of Devonshire kept to her promise to canvass support, beginning with 100 guineas from her husband.

If Beddoes was constitutionally incapable of harnessing all his energies to a single task, his extramural activities at least began to slow to a comparative trickle. He accepted a commission, politely turned down by Darwin, to edit a new edition of John Brown's *Elements of Medicine*: a charitable initiative for the benefit of Brown's recently bereaved and debt-ridden family, which also offered a chance to address the nagging question of how Beddoes' pneumatic scheme stood in relation to a Brunonian system that time was not treating kindly. He had also begun to write anonymous pieces for the liberal and learned *Monthly Review*: commentaries on animal electricity and the new chemical nomenclature, crisp summaries of German monographs on yellow-fever epidemics and the physiology of respiration. But these side projects had a sharper focus than the effusions of two summers before: each provided context for his main project, or kept up the drip-feed of scientific progress to the interested public who would make up his subscribers. Beddoes' restless years were firmly behind him; he had finally found a cause into which all his profuse talents and energies could be poured.

But the work schedule demanded by this cause was punishing. Over the summer of 1794 Beddoes travelled constantly, visiting prospective backers and explaining his scheme to doctors, chemists and industrialists. Anna, left behind in Clifton, felt the withdrawal of attention keenly, though she was still able to express it playfully. She wrote to him during a foray to Shropshire in July with the good-humoured but rueful prediction that 'this letter will not be opened with the impatience that it would have been some little time ago, yet I am very sure you will receive it with pleasure'. She wished to tell her husband 'all that I think and feel, it is impossible for you to tell what real happiness this gives me, because since you have taught me to love you so much, it would be very hard if I might not tell you so'; but she was conscious that such talk 'is out of season with you'.[55] She had, she was now well aware, married a profoundly good man who adored her, but whose priorities and passions would always leave her with less of him than she needed.

On these travels Beddoes was beginning to circulate a four-sheet handbill that formally proposed, and titled, his project for the first time. 'A Proposal towards the Improvement of Medicine' announced the establishment of a Medical Pneumatic Institution, to be funded by charitable subscription. Since it had now been 'abundantly proved, that the application of elastic fluids to the cure of diseases, is both practicable and promising',[56] the logical next step was to determine by experiment the best method of administering the gases and their effects on various diseases. Beddoes was certain that this would 'be much more speedily accomplished by means of a small appropriated Institution, than in twenty years of private practice', and that funds should be equally speedily raised 'since nothing is more urgent than to restore health, and preserve life'. The costs of the project were enumerated – premises; chemical equipment and materials; a superintendent, along with with two assistants and a nurse; medicines and administration – but the precise sums were, like the fluids, elastic: the budget 'might be expanded or contracted, according to the amount of the contributions', although 'three or four thousand pounds would probably suffice'.[57] A list of private banks in London, headed by Coutts, was already standing by to accept either private subscriptions or transfers from provincial bankers.

It was Tom Wedgwood who came forward to place himself at the head of Beddoes' supporters. In August, Beddoes wrote to his father, Josiah, with whom he was already in correspondence about subscribing to the Institution, assuring him that 'I have reason to believe that oxygene air will be almost infallible in putrid fevers as they are called', and that the duchess of Devonshire was also taking up his cause 'with much ardour'. Tom scribbled encouragement across the top of his father's correspondence: 'I think I shall contribute as the attempt must be successful in part if it only goes to prove that the airs are *not* efficatious in medicine. And he seems determined to give them a thorough trial.'[58] If Tom was happy to drum up a subscription from his father, he was also keen to play down his expectations. His father's search for a cure for his condition had been as intensive, and often as agonising, as Watt's exertions on Jessie's behalf, with thus far as little success and no guarantee against the same tragic conclusion. With characteristic delicacy, Tom was seeking to distance himself from the hope of a miracle cure which, if it became an expectation, would place a cruel stress on Josiah, and perhaps a crueller one on Beddoes himself.

By the end of the summer a thick manifesto for the new Institution was rolling through the press, entitled *Considerations on the Medical Use of Factitious Airs, and on the Manner of Obtaining Them in Large Quantities*. It was presented as a medical treatise in Beddoes' ongoing series, but its tone had unmistakably shifted towards a work of advertisement and advocacy for the project it was designed to launch. Its preface was the 'Proposal towards the Improvement of Medicine', after which it began with first principles – normal air is a little over a quarter oxygen, the rest nitrogen – and with simple experiments that could be tried at home: 'if you fix a pipe to a bladder full of air, and holding your nostrils breathe this air for some time, your distressed feelings will inform you that it is no longer fit for breathing'.[59] It was a work clearly less interested in impressing doctors with its learning, and more obviously concerned to draw in a readership of potential charitable donors.

From such simple beginnings, *Considerations* proceeded to lay out the findings of Beddoes' experiments with kittens and other animals in Hope Square the previous year. He had measured the quantity of oxygen expended by kittens in a bell jar under various conditions (normal, harassed, agitated, drunk on sherry), and had timed the survival of rabbits in different airs (dead in two minutes in carbonic air, in under five in hydrocarbonate, still alive after ten minutes in hydrogen). The findings are carefully, perhaps coldly, tabulated, and Beddoes acknowledges that the experiments had their grim moments – 'in a few cases, the torture which was inflicted was exceedingly repugnant to my feelings' – but he offers the reader no apology. 'I know not', he maintains, 'how he who feeds upon the flesh of a slaughtered animal can, upon reflection, condemn investigations seriously tending to restore or preserve health'.[60]

In contrast to his previous publications, Beddoes' theories about the effects of factitious airs set out here are brief and tentative. An atmosphere containing reduced amounts of oxygen may have 'soporific virtue', as well as preventing or ameliorating consumption;[61] one rich in oxygen may be effective against asthma; but, as in his letter to Watt, he does not wish the feasibility of his project to be judged on his earlier and more sweeping predictions. What he is calling for is simply 'a trial of the airs in medicine, without giving the smallest assurance of success', and 'in whatever cases the practice proves useless or disadvantageous, I shall as earnestly dissuade from it as I before advised the trial'. Perhaps under the steadying influence of Giddy, or the confidence instilled by recruiting James Watt, he was recalibrating the balance between theory and experiment, preferring an empirical programme of trial and error to the alternative of making himself a hostage to claims that might be quickly exploded. His most powerful argument, he had decided, was the one with which Watt had explained his involvement to Darwin: he might not know the answers to his questions, but assuredly no one else did either.

The second part of the *Considerations* was turned over to James Watt in person, his Bastille Day letter to Beddoes reproduced along with his elegant diagrams and subsequent correspondence, all stamped with a handsome facsimile of his personal signature. Although Beddoes assures the reader that 'to procure a dose of factitious air by means of Mr. Watt's apparatus will, I think, be found quite as easy as to dress a joint of meat',[62] the chemistry required to produce the airs is still plainly complex and dangerous: reactions with iron and zinc will become red-hot, all charcoal must be thoroughly burned in advance to avoid inhalation of toxic smoke, and every joint in the apparatus must be carefully sealed to prevent leaks and explosions. There were certainly many private chemistry laboratories across the country where such operations could be easily performed, yet the claim that they could be rolled out to anyone who had a kitchen stove was an ambitious one. In subsequent editions Watt would add a caution that even a naked candle flame might cause the entire apparatus to explode.

Beddoes concluded *Considerations* with a note entitled 'Hyper-Criticism', directed at a less welcome but still necessary constituency: the book's potential reviewers. 'You have heard the project vilified', he tells them. 'So would a panacea be. So was the Peruvian bark; and inoculation; and every great improvement of that art, from which, according to its state, all in their turn shall experience good or harm.' But this is too important an issue to be derailed by the fashionable mocking of novelty, the urge to debunk innovation in general or to belittle provincial outsiders. Let reviewers remember the probability that they will find themselves on a sickbed soon enough, and consider whether they would then still cherish their witticisms that assured the public there was no need to take pneumatic therapy seriously.

Through the autumn the drive to publicise and fund the Institution continued at a hectic pace. Beddoes' letters to Tom Wedgwood record the growing list of luminaries prepared to lend their names to the project – Joseph Black, Erasmus Darwin, James Watt and Matthew Boulton, Josiah Wedgwood, James Keir, the duke and duchess of Devonshire – and those whose endorsement he is still pursuing, such as the reclusive chemist Henry Cavendish and, most crucially, the president of the Royal Society, Sir Joseph Banks. 'I have within a few days written scarce less than a hundred letters on the subject of the Institution',[63] he reports. He was licensing 'agents' – supportive public figures to canvass local physicians, press and public – from Chester to Yarmouth, Liverpool to Exeter, Shrewsbury to Southampton to Cornwall; Erasmus Darwin had opened a subscription in Derby, and had already raised 32 guineas. Hospitals in Bath and Manchester had begun trials with factitious airs, and were asking Beddoes for technical advice. The Medical Pneumatic Institution was no longer a notion to be speculated upon, but a reality to be engaged with.

Meanwhile, world events continued their explosive chain reaction but, for once, without Beddoes' commentary. Danton had been executed, and the Terror had hurtled to its bloody conclusion with the grotesque death of Robespierre. Under the martial discipline of the new Thermidorean regime, order was being restored, the Jacobin Club closed down and the *sans-culotte* mob chased from the streets of Paris by the resurgent royalists and their strut- ting *jeunesse dorée*. In London the government's landmark treason trial against Thomas Hardy, John Thelwall and their alleged fellow-conspirators had collapsed, leaving the attempt to redefine dissent as sedition in disarray. Both British and French governments, blinded by their intransigent hatred of one another, seemed to have derailed themselves with destructive and divisive excesses. Yet even Beddoes' letters to Giddy, so often the clearing-house for his political thoughts, had shrunk down to dashed-off summaries of his progress with fundraising, the 'hope that there are persons enough in the kingdom who are humane' and the question of 'how much is a handsome subscription'.[64] There was much else for Beddoes to say, but not a spare minute in which to say it.

* * *

As the Medical Pneumatic Institution rolled its campaign out across the country, its strategic weakness became ever more obvious: it was spreading effectively through the provincial networks where it had its core support, and making inroads in several new ones, but making barely any impression on the metropolis. London was where the densest concentration of potential subscribers was to be found, and it was also the area where Beddoes' sympa- thisers were fewest and least visible. But the concentration of the scientific establishment in London also meant that the right imprimatur might secure it wholesale at a stroke. The endorsement that would do more than any other to deliver the city was that of the Royal Society, and the Royal Society was effectively the fiefdom of one man, Sir Joseph Banks, who had been its president since 1778 and whose personal network had come to define London's scientific and medical elite. It was not even necessary to ask the Society for funding: if only Banks were to allow the Institution to operate under his aegis and with his approval, it would open the doors to the nation's largest and wealthiest constituency of doctors, researchers and charitable donors.

Beddoes was by no means unknown to Banks, who had published two of his metallurgical papers in the Society's *Proceedings* only three years previously; now, he judged, he had two plausible conduits to him, both of whom began manoeuvres to enlist his support in November. One was Georgiana, duchess of Devonshire, who was well acquainted with Banks from salons on the Continent and Royal Academy lectures in London; the other was James Watt junior, now twenty-five and well established both in science and in business. James had, on his excursion to Paris with Tom Wedgwood, been travelling as a textile salesman, a profession in which had since become highly successful; he was also, like Tom, a gifted chemist, and had over the years met and conversed with most of the French authorities on the subject, including Antoine Lavoisier. He had begun selling Beddoes' scheme among his friends, and had already secured subscriptions from his father, Matthew Boulton and the Lunar doctor William Withering, and was working his way through a long list of Birmingham physicians.

It was Georgiana who took Beddoes' scheme to Banks first, and who received the more considered response. He was concerned, he began, that Beddoes was 'a man who has openly avowed opinions utterly inimical in the extreme to the present arrangement of the order of society in this country'; and although Georgiana's support of the project had softened his feelings on that score, he was equally troubled by the risks involved in the proposed experiments, fearing that 'there is a greater probability of a waste of human health, if not of life, being the consequences of the experiment, than an improvement in the art of medicine being derived from the results'.[65] To risk the life of innocent subjects for speculative and perhaps illusory benefits was a plan for which, with the deepest regret, he felt unable to offer his support. Georgiana pressed him, assuring him that Beddoes 'had long abjured' his erstwhile political opinions and 'was wholly given up to his philosophical experiments', and adding that 'I think your protection of such consequence to any undertaking that I shall feel inclined to advise

Dr. B to give up any public trial without it'. Banks shrugged off this implied obligation with the assurance that 'he cannot suppose his name a matter of material consequence to Dr. Beddoes', and expanded his objections to include the well-known risks involved in extrapolating from animal trials to human subjects.[66]

The young James Watt, even though he was in regular contact with Banks on other matters, received even shorter shrift: another insistence that 'you flatter me by saying that my name will be of use to Dr. Beddoes', and a request 'that I may not be pressed any more by the doctor's friends to do what I have already ... formally declined to do'.[67] Watt was stung, and unimpressed by Banks' pretensions to concerns about scientific method, especially since 'all the men of real chemical knowledge' in the country had already lent their names to the plan. These were pretexts, in his view, for an anxiety that was nothing to do with experiment or evidence. 'The fact is I suppose', he wrote to a friend, 'he has seen Beddoes' cloven Jacobin foot and it is the order of the day to suppress or oppose all Jacobin innovations such as this is already called'.[68]

It was a tangled outcome, not unlike Beddoes' departure from Oxford, and probably in part a consequence of it. There, his resignation had technically preceded his pamphlet on the September Massacres, but the two had been conflated. Now, in addition to his scientific objections, Banks had clearly been swayed by Beddoes' political opinions, and probably by his inclusion on the list of political undesirables that had been circulated by the Home Office in the wake of his Oxford resignation. But Banks had other grounds for caution. His imprimatur was a powerful tool for setting projects in motion precisely because he was so cautious in granting it; and if he had been paying attention to Beddoes' endeavours, he would likely have noticed his tendency to make free with the names of his distinguished supporters such as Erasmus Darwin, often imputing to them views they did not necessarily have. Beddoes was a fine advocate for the benefits of innovation, but less eloquent about its dangers; if the Royal Society were to endorse his scheme, Banks was well aware that he would share the risks as conspicuously as the rewards.

The figures who approached Banks on Beddoes' behalf were also less impeccable than they may have appeared from Beddoes' provincial perspective. Georgiana was assuredly still a name to conjure with, but her endorsement was no guarantee of quality: she was also known for her patronage of James Graham, the most flamboyant quack of the day, at whose 'Temple of Hymen' in Pall Mall couples could hire a 'Celestial or Electro-Magnetical Bed' that guaranteed fertility and perfect progeny, and was available for £50 a night. Would the duchess expect Banks to endorse Graham's array of nostrums that ranged from 'Electrical Aether' to an 'Elixir of Life'? And if not, could she explain how precisely they were different from Beddoes' factitious airs? And though James Watt was the son of the nation's greatest inventor, and a respectable businessman and fine chemist in his own right, he was still known above all for his politics: in 1792 he had even briefly been a delegate at the Jacobin Club in Paris, and Edmund Burke had singled him out in the House of Commons as a rare and conspicuous British ally of that society of regicides. During the September Massacres he had witnessed the body of the princesse de Lamballe being dragged naked through the streets, abused and beheaded; though he had been 'filled with involuntary horror' at the scenes, he neverthe-less insisted on 'the absolute necessity of them'.[69] Though he had become more discreet about his politics in the meantime, he had not exactly recanted; if Banks was concerned that Beddoes' zeal for innovation and political reform might shade into support for an enemy in time of war, James was not the man to convince him otherwise.

For Beddoes, who had spent the year convincing the world, and himself, that the Institution promised discoveries on a spectacular scale, it was a harsh colli-sion with reality; and it would not be his last. By reaching out to publicise and fund his project, he was also attracting the attention of his detractors, and along with support and subscriptions came lampoon and hostility on a level he had never previously experienced. At the end of 1794, an anonymous pamphlet entitled *The Golden Age* and purporting to be 'A Poetical Epistle from Erasmus D—n MD to Thomas Beddoes MD' emerged from Oxford. Its two hundred lines of heroic couplets were crudely modelled on Darwin's epic poetry, and possibly on Beddoes' *Alexander's Expedition*: some choice lines from his pamphlet on the September Massacres were reprinted on the title-page, suggesting that the anonymous author had been paying attention to Beddoes' output from Shropshire two years before.

The other introductory quotation, which *The Golden Age* took as its text, was a rhetorical question from *Observations on the Nature and Cure of Calculus* where Beddoes, speculating on the vistas of progress that might be opened up by science, had asked, 'May we not, by regulating the vegetable functions, teach our woods and hedges to supply us with butter and tallow?' In this vein, 'Darwin' eulogises Beddoes as the spirit of the age of wonders that are destined to spring from his ceaseless quest for innovation. He is the 'boast of proud Shropshire, Oxford's lasting shame'; in politics 'the bigot's scourge, of democ-rats the pride'; and, in his chemical quackery, the 'Paracelsus of this wondrous age'. Darwin conjures a future – 'Things strange to tell! Incredible, but true!' –

where sheep will be bred to be purple and scarlet, the dairymaid will abandon her 'weary churn' and watch contentedly as pounds of butter drop from the hedges, and even the river Thames will

> Resign his majesty of mud, and stream
> O'er strawberry beds in deluges of cream![70]

The Golden Age is a lampoon of the broadest sort and yet, with hindsight, it is remarkable how many of its facetious prophecies now seem uncannily prescient. We have long substituted vegetable oils for butter, releasing the dairymaid from her weary churn; now, a vast biotechnology industry promises a similar transformation of vegetables into factories of nutrition and medicine. As Darwin's journey continues through cloud-cuckoo-land, he witnesses a future where 'happy man / to ages shall extend life's lengthened span', and the world will be populated with healthy and attractive eighty-year-olds. In this fantastical world, belief in Christianity will have waned, and time-honoured social institutions will be no more ('see tythes expire, and ancient slavery fail!'). Beddoes' conviction that chemistry was poised to remake the world in ways his contemporaries had yet to grasp is strikingly illuminated by this spoof on the absurdity of his predictions.

Yet however wildly it may have misjudged its targets, *The Golden Age* was grimly instructive to both the objects of its parody. To Darwin, it demonstrated that the cosmic materialism and implicit atheism of *The Loves of the Plants* were now out of season, and that his recent friendship with Beddoes had left him exposed. No longer could he expect to charm the reading public with urbane speculations that sex, rather than religion, was the true motive power behind the living world: the line of tolerance that had separated him from his more politically engaged associates such as Joseph Priestley had now been erased. Darwin took the point of this and similar lampoons, and became scrupulously careful to give religion and politics a wide berth in his future writings.

To Beddoes, the attack demonstrated that he would, from now on, be forced to play a nearly impossible hand. He could not recant his belief that science was on the cusp of transforming the world in ways that even he could barely imagine, and that in his chosen field – the application of chemistry to medicine – the transformation was imminent, and unprecedented in scope. But the very scale and novelty of the revolution he was pursuing made it profoundly suspect in the new climate, no matter what benefits it promised to deliver. He ended 1794 with the wind in his sails, but a storm blowing up behind his back.

4

The Watchmen

The new year began with the coldest winter in fifty years. In Paris, as the military battened down on the frozen streets and Robespierre's allies in the Terror were thinned by convulsive purges, the Seine stayed frozen for weeks, and wolves began to creep down from the hills to scavenge at the edge of town. In Britain the third poor harvest in succession drove food prices to twice their 1792 levels and, as Pitt's war taxes and army requisitions ratcheted up shortages to starvation levels, the hubbub of voices calling for peace began daily to test the strictures against sedition that had foundered so publicly at Thomas Hardy's trial. In Cornwall, Davies Giddy found that mobs of miners and farm labourers could no longer be dispersed with a few bushels of barley; beneath the complaints of hunger was a new and implacable distrust, a 'general fermentation among the labouring class',[1] where the shadow of violence now lurked behind every negotiation. Giddy began to take urgent stock of his county's grain supplies, setting up rationing and commandeering imports from the American colonies.

For Beddoes and his circle, the beginning of 1795 was marked by personal tragedy. Josiah Wedgwood, who had in recent years suffered from intermittent pains in the jaw, collapsed over Christmas into a fever, his face swelling with infection. Darwin, in many ways his closest friend as well as his long-time physician, had told him breezily in mid-December that he was on the mend, but by New Year he was nursing him through a helpless decline and on 3 January pronounced him dead. Josiah's sons, John, Jos and Tom, now stood, ready or not, at the head of one of the nation's leading manufacturing dynasties. For

Tom, who had been devoted to his father, it was a crushing blow. The responsibilities and anxieties that had led to his breakdown were intensified; decisions pressed upon him just as he was least able to make them. Along with his crippling illness he was now, at the age of twenty-three, a man of considerable wealth. Should he remain in the family seat of Etruria, or make a new start? Should he travel, or retreat deeper into his bed? And what other possibilities, thus far unimagined, might his inheritance open up to him? He turned increasingly to the stolid, calm figure of his brother Jos, and to Beddoes' medical care.

Despite the setbacks of the previous year's end, the Pneumatic Institute, as it was becoming known, was building up a powerful head of steam, and burying Beddoes in 'an immense correspondence which takes my whole spare time'.[2] Subscriptions had begun to arrive – 'I suppose £500 or so' – although he had delayed placing advertisements in the newspapers, feeling it indecent to ask for funds 'from the necessity of contributions to keep the poor alive this hard winter'.[3] With the spring thaw, the advertisements began to run, making their debut in the Birmingham press, and his exertions seemed finally to have given the project critical momentum: as he put it to Giddy in the phrase of the French Revolution's uplifting anthem, 'I begin to think of the Pneumatic Institute scheme *Ça Ira*'.[4] But he was anxious that if the war, shortages and bankruptcies worsened, no such venture could be sustained. 'I think the common people are getting dissatisfied', he wrote to Tom Wedgwood; and 'reflections like these make me apprehend that the Pneumatic Institution will be defeated by public disasters, even if the subscription should be adequate'. But for the time being at least his spirits burned brighter than his doubts: 'this apprehension is only with me a reason to try and get it forward'.[5]

But while the project was making a powerful impression on the physicians of Bristol and the Midlands, in other areas professional inertia was harder to shift. Although trials of pneumatic medicine were now under way in several provincial cities, Beddoes was still unable to find a single London physician prepared even to make a test of the airs. He had plans for a superintending committee, but to establish one without a single London name of note would only draw attention to the lack; he was eventually forced to concede that, among the 'eminent doctors' of the metropolis, 'I know not one who has such an enlarged understanding, much less who has any uncommon knowledge of physiology or chemistry' to grasp the point of the project.[6] Giddy, too, was drawing a blank: 'I am really sorry for the honour of Cornwall that not a single subscriber except myself is likely to be found in the whole county'. It was not, in his view, the project itself so much as the novel proposition of charitable donation to a scientific cause rather than to a 'visible object of distress'. Nevertheless, Giddy himself remained a solid supporter of the project, and continued to add suggestions to the range of therapies it might offer. If factitious air had the power to calm excitability, he inquired of Beddoes in a prescient afterthought, 'might it not be used before painful operations?'[7]

With Giddy confined to Cornwall and weighed down with matters of law and order, it was Tom Wedgwood who now became Beddoes' closest confidant in the project. His father's death had brought on a prolonged and severe bout of the mysterious stomach complaint that had periodically crippled him since childhood, but Beddoes' treatments, particularly charcoal, had relieved it; now Tom, awakening to a world where he was no longer the troubled heir but the master and patron, began to involve himself closely in the Institution's plans. He and Beddoes discussed 'a pamphlet for readers of fashion' to introduce the idea of pneumatic therapy to the metropolitan salons, featuring 'a frontispiece with Mr. Watt's apparatus and people at work, a vignette with a patient inhaling from a bag and a tail-piece representing a breathing apparatus'.[8] For both, these exchanges were an expression of a warm fellowship, behind which the balance of give and take was exquisitely intertwined. On one level, Beddoes was performing the service of drawing Tom out of his sickbed and into the world of business, perhaps quite consciously as part of his therapy; on another, he was easing Tom towards a deeper level of commitment that would in time also become a financial one. Tom, in turn, was milking Beddoes for experience in coping with his new role as patron, confident that the price tag for the lesson was one he would be glad to pay. And behind all these transactions remained the bedrock of their relationship as doctor and patient. Tom, as his strength returned, became keen to undergo oxygen therapy himself; Beddoes was delighted by his enthusiasm, though he cautioned him to stick to moderate doses. They discussed the accounts and case histories that Beddoes was now receiving almost daily from physicians and patients, and Tom began writing up his own case notes for publication.

By March, Beddoes had a new edition of *Considerations on the Medical Use of Factitious Airs* in the press: 'nearly a new book', he announced to Giddy, with 'only 50 out of 220 pp.' the same as the prior edition.[9] Parts one and two had been corrected and extended, and a third part had been added, with 150 pages of letters and case histories, including material from familiar allies like Darwin but also around twenty new recruits to the cause, recording all the experiments, including several apparent cures, that had been made since the previous autumn. Beddoes also included an introductory 'Epistle General', a dramatic sketch of broad satirical intent: an odd inclusion in a medical text, but one that revealed the extent to which the Royal Society's rejection, and in particular the lampoon of *The Golden Age*, had got under his skin. This was satire from the opposite ideological pole, in the form of a dialogue between two doddering octogenarian physicians lamenting 'the degradation of our art' since the idea of progress and experiment had taken hold in medicine. No longer did their Augustan wigs and canes command unthinking deference: apothecaries were cutting into their trade with newfangled preparations, and impertinent young doctors were abandoning the dignity of Latin and addressing their patients in a comprehensible vernacular. Worst of all, a new generation of 'speculatists, enthusiasts and high-flyers' had taken to experimenting on their patients. One doctor tells the other of his patient, the consumptive daughter of a lord, whom he has informed that 'the disease is always fatal'; nevertheless, the lord wishes her to be sent to Hotwells for a possible cure. The physician consents reluctantly, but assures the girl's father, 'I will answer for it, no *experiment* is performed on the patient'. To those who offer troublesome evidence that pneumatic therapy is already beginning to rescue these hopeless cases, the ancient physician declaims the mocking challenge:

> Proceed, high-toned enthusiast! Coax mankind
> With idiot mouth to gape and suck the wind![10]

* * *

Beddoes had, for two years, followed Edgeworth's prescription to restrain himself from politics until his medical ambitions were fulfilled; but by the beginning of 1795 it began to seem that politics was the most serious obstacle in his path, and one that could not be ignored for much longer. The frosts thawed, the rivers flooded and food supplies began to move once more; but it was clear to him that the threat of social collapse had only been temporarily stayed, and that 'nothing but a peace which I much doubt whether we can obtain can save this country from universal distress'.[11] Britain had declared war in 1793 against a revolutionary rabble: the government had expected a swift campaign, with overwhelming support from the Continent's encircling monarchies, after which ripe colonial plums such as Santo Domingo and Ceylon would fall into its lap. But the revolutionary rabble had become an inspired fighting machine, and now a new and terrifying martial state; the Continental powers had been beaten back, and Britain herself was starving and bankrupt. It was ever less plausible that Britain was fighting a war of self-defence against Jacobinism, and ever clearer that she had joined the unreconstructed despots of

Europe in a predatory grab for wealth, territory and power. The war had lost its sense, and its future was now all too predictable: an expensive stalemate, with potential gains far outweighed by crippling costs, each side grinding against the other until one finally collapsed in exhaustion.

Beddoes' concerns were widely shared. In the aftermath of the winter shortages, campaigns and petitions for peace began to mount: thousands of working men, politicised and emboldened by the collapse of Hardy's treason trial, began to sign up for the Corresponding Societies in London and the provinces. When a petition of citizens in favour of peace was mooted in Bristol, Beddoes joined the organising committee, asking the mayor, James Harvey, to chair a meeting in the Guildhall. Harvey 'at first seemed eager', until he began to receive discouraging soundings from high places: by Beddoes' account, particularly 'when practised upon by some friends of the Duke of Portland',[12] the leader of the Whig faction which had recently distanced itself from Charles James Fox's calls for peace and joined Pitt's administration to support an even more vigorous prosecution of the war. The mayor withdrew his support, and refused to offer the Guildhall as a meeting place in his absence.

In the course of this agitation Beddoes came into contact with a new generation of political activists who were determined to supply the voice of protest that had traditionally been so muted among Bristol's citizens. Conspicuous among them was the twenty-four-year-old Robert Lovell, son of a wealthy Quaker family with manufacturing interests. Though the Quakers were a powerful economic force in Bristol, they, like the other dissenting communities, had long been in a delicate position in a city whose wealth depended so heavily on the slave trade. Since 1758 the Quaker Society of Friends had forbidden its members from taking any part in the 'iniquitous practice' of slavery, the first formal embargo on the trade in British civil society; but their protests had been peaceful, boycotting slave-made products, setting up fair-trade networks and encouraging alternative ethical trades, such as those in East Indian sugar and Canadian maple syrup. But Lovell, coming of age in the new confrontational climate, was not content to protest as discreetly as his parents had done, and he had taken to poetry to skewer the hypocrisies of Bristol and its mercenary culture. In his *Bristol: A Satire* he sketched a savage portrait of a city where

Trade, mighty trade, here holds restless sway
And drives the nobler cares of mind away
. . .

'Tis 'How goes sugar? What's the price of rum?
What ships arrived? And how are stocks today?'[13]

Lovell's dedication slyly invoked the Bristol Bridge riots in support of his satire, but his was a different voice from that of the turnpike rioters, and from that of the many crowd disturbances and strikes of working men that had preceded them. It was a wilder howl of protest against a culture lorded over by a sluggish and corrupt elite: both a local Corporation bloated on the profits of a callous trade that coarsened the souls of those who worked it, and a national government that staggered on with a war motivated by colonial greed, suppressing the voices and liberties of its subjects in the process. It was a voice that chimed with Beddoes' insistence that trade, if unfettered, fed the predation of the weak by the strong in ways that made a nation sick: wealth, accumulated for its own sake and hoarded alongside the privation and misery of others, led to a distorted body politic and a diseased society.

Among Lovell's circle was another poet, political reformer and son of a wealthy Bristol family, Robert Southey. An earnest and intense Oxford student three years younger than Lovell, he was expected by his family to take holy orders but had become a zealous and unrepentant supporter of the French Revolution: he had cheered on the Terror and regarded Robespierre's fall as 'the greatest misfortune Europe could have sustained'.[14] Southey shared Lovell's disgust with Bristol, and together the two were planning their escape. They would assemble a company of like-minded republicans and emigrate 'to the banks of the Susquehannah' in America, where they would form their own independent society, far from the madness of the bellicose tyrants who were set on pounding Europe to rubble. Lovell had already found his partner, a local girl named Mary Fricker of whom his parents strongly disapproved on the grounds that she was not only a non-Quaker but an actress, and thus by their creed a professional teller of lies. Southey would soon be courting Mary's sister Edith; and before long they would be joined by Samuel Taylor Coleridge, who would fall in love – or so he believed at the time – with the third Fricker sister, Sara, join their American plan and christen its philosophical scheme 'Pantisocracy'.

Both Southey and Coleridge, as they aged, grew in stature and renounced (or simply denied) the politics of their youth, would give the impression that Pantisocracy and their utopian society on the Susquehannah were juvenile dalliances, madly impractical and ultimately a frivolous flight from reality. Yet in its essence their scheme was a sane response to the times, and indeed one that was quite commonplace: hundreds of people a year were leaving Britain for a new life in the wide open spaces of America, and particularly for new states such as Pennsylvania where a strong Quaker influence had created a haven of tolerance for dissenters and freethinkers. The direct inspiration of the Pantisocrats was Joseph Priestley who, two years previously, had slipped away from Hackney to set up home and laboratory in Northampton, Pennsylvania: they conjured him up in their imaginations, living the natural and honest life that he had always preached in a rustic log cabin surrounded by virgin hills, far from the restentments and factions that had brought him so close to being publicly butchered. Emigration was often in Beddoes' thoughts, too, and Davies Giddy, forced into such detestable choices at home, was also seriously considering it, and had even been putting his money into American stocks. 'You would find yourself more at home in Philadelphia than Tredrea', Beddoes had urged him at the end of 1793; 'you might not only gain respect but money with your knowledge there'.[15] With England and France locked in a grim and seemingly interminable conflict, America was far more than an idle fantasy for utopian dreamers.

It was shortly after the arrival of Coleridge, who moved into shared lodgings in the centre of town with Lovell and Southey in early 1795, that Beddoes probably first met the Pantisocrats, and his friendship with Coleridge in particular developed into a communion of rare scope and intensity. A year out of Cambridge and fresh from his quixotic adventures as an army volunteer, Coleridge presented a wild spectacle, black curls of hair falling untidily over the collar of his brass-buttoned army greatcoat and invariably in the throes of passionate speech. In the words of Joseph Cottle, the bookseller in whose shop on Bristol High Street the two may have met, 'during the whole of his residence in Bristol, there was, in the strict sense, little of the true, interchangeable conversation with Mr. C. Almost, on whatever subject he essayed to speak, he began an impassioned harangue of a quarter, or even half an hour.' Coleridge may occasionally have given way to Beddoes' own Shandean rambles from chemistry to philosophy and via ancient civilisations to the politics of the future, though even 'inveterate talkers', according to Cottle, 'generally suspended their own flight' in Coleridge's company, feeling it 'almost a profanation to interrupt so mellifluous a speaker'.[16] Yet Coleridge would absorb a great deal from Beddoes in the course of the monologue that was his waking life, and the friendship they formed would, most unusually for Coleridge, remain warm and unbroken for a lifetime.

For the Pantisocrats, whose world consisted predominantly of other young men, mostly still at university, with their strident critique of everything that was rotten in the world around them but only a nebulous sense of an alternative, Beddoes was an inspiration. He had forged his way through the world without abandoning his principles; he was both a rebel and a success. 'The faculty dislike Beddoes because he is more able, and more successful, and more celebrated, than themselves', Southey recorded triumphantly.[17] His reputation as a political firebrand was to them admirable, founded as it was on the same passion for liberty and equality as Pantisocracy, although even in his youth Beddoes had been careful to draw a clear line safely on the moderate side of mobs and blood-shed. His learning, and the vast spectrum of disciplines and languages that supported it, was on a scale that even the omnivorous Coleridge could not yet encompass. Beddoes was also an intimate of Erasmus Darwin, whose poetry represented the summit of the art to which they aspired, and he stood as the inheritor of the mantle of Joseph Priestley, their inspiration and idol.

He also had a very congenial town house in Clifton, a far more agreeable base than their cramped lodgings in dingy, noisy College Street, with an Aladdin's cave of a library containing well over a thousand books, many in French and German, including shelves of foreign journals that explored new theories and philosophies barely known in Britain. The young Coleridge was, by his own description, a 'library cormorant' who gobbled down every book he could lay his hands on, and Rodney Place was a treasure that both Beddoes and Anna were delighted to share. Beddoes had given up Oxford, but he had not lost his urge to pass on his learning to the next generation, and here was an eager class of the most able students he had ever met, ravenous for all he had to feed them. For Anna, whose greatest dread was an empty house with her husband constantly absorbed in business, it was a relief to fill it with the chatter of a group much closer to her in age than her husband. Both Beddoes and Anna were informal, delighted by spontaneous visits and readings from works-in-progress, and Rodney Place soon hummed with the buzz of a literary salon.

There was an endless range of subjects on which Beddoes' and Coleridge's interests coincided, and around which their monologues danced and interwove. Medicine, the bedrock of Beddoes' intellectual world, was also for Coleridge the most significant of the sciences, the key to the mysteries of sensation and perception, the formation of ideas and the relationship between the body and the mind. Like Beddoes, he was fascinated by the Brunonian system: its central idea of excitability implied a store of vital energy expended over the course of a lifetime, and suggested that philosophical ideas such as mind and spirit might eventually be made subject to physical analysis and experiment. Coleridge had likewise absorbed the work of David Hartley, whose vision of the brain and nerves physically forming associative channels in our tissue underpinned the principles of Coleridge's harmonious Pantisocratic community, just as it did Beddoes' theories of education. As their conversations ranged across the borders between the physical and the mental, and the grey areas between reality and idea, they discussed the physical causes of hallucinations and analysed their dreams.[18]

But there was one particular field of study in which Beddoes was an expert teacher and Coleridge an avid pupil. A few months previously, in November 1794, Coleridge had sat down one midnight in Cambridge to read the recent translation of Friedrich Schiller's first play, *The Robbers*, in which the protagonist abandons the corrupt world of money and power for an outlaw utopia in the forests of Bohemia. It was his first encounter with the new German literature, and it had left him stunned. 'My God, Southey', he wrote to his friend, 'who is this Schiller, this convulser of the heart?'[19] Now, at Rodney Place, he found himself in perhaps the finest private German library in the country, where the medical works of Christoph Girtanner and Göttingen University's Brunonians sat alongside the works of Goethe, the drama of Schiller beside the sceptical philosophy of G.E. Lessing, and scholarly journals containing many of Beddoes' own reviews and translations. Beddoes had, for example, followed closely the new biblical 'higher criticism' being developed in Germany by scholars such as J.D. Michaelis and his student and successor J.G. Eichhorn, for whom the Bible was not a sacred repository of revealed truths but a collection of texts of near-eastern Bronze Age mythology, a companion piece to the Hindu literature that Beddoes had mined for *Alexander's Expedition*. For a Unitarian such as Coleridge, these works were fascinating for a different reason: they offered the chance to reconstruct an original and universal Christianity, purged of the political machinations and corruptions of Rome.

In the course of their explorations in German literature, Beddoes and Coleridge developed a growing interest in the thought of Immanuel Kant. Beddoes was revising his earlier dismissal of Kant, and preparing a long summary of his political philosophy that would be published in the *Monthly Review* the following year. On first encounter he had taken Kant's claim that certain truths could be known *a priori* to be a reactionary stance: an attack on reason and experiment, and a defence of faith and revealed truths. Now, however, he recognised that Kant was in fact attempting to push the sceptical ideas of Locke and Hume to their logical conclusion. He was agreeing that all we know of the world is derived from the sense-impressions that we receive, and that if there is a 'real' world beyond these we can know nothing of it. But he was also plotting a route through this impasse by arguing that the mind not only receives these impressions, but also orders them and structures them in its own image. The world seems the way it does to us because that is the way the mind works: it has its own *a priori* categories, such as space and time, within which eveything we know is necessarily situated. Kant was not arguing that space and time are absolutes, but that they are an inevitable part of the human experience. Consequently, if philosophy and science wish to progress, they must do so by investigating the structure of the mind, for it is there that the deepest reality lies.

Unlike Beddoes, Coleridge had an unquenchable appetite for metaphysics and specifically for worrying at the relation between personal experience and universal truth; and, as with the new biblical criticism, he took from Kant something quite different from what Beddoes did. There was a second component of Kant's philosophy that Beddoes dismissed, but that came to assume a central importance in Coleridge's thinking: that, although it was strictly true that we have no knowledge of any ultimate reality, we must nevertheless believe in God, because doing so gives us the possibility of acting in ways that are morally correct. In this sense Kant's work – like Hartley's, which also proceeded from material causes to religious conclusions – both stimulated the intellectual enquiries that were common to both Beddoes and Coleridge, and exposed the glaring faultline that ran between them: religion. Beddoes was interested in the origins of religion and the forms it took in different societies, but he was never troubled by questions of faith: like Erasmus Darwin, he found that the more he understood of the world, the less he could be bothered with theology. Since the September Massacres he had seen little to change his view that religious belief was more often than not a source of philosophical error, bigotry, prejudice and corruption, and an enemy of rational progress towards a civilised society.

Coleridge, by contrast, was unable to conceive of a world without faith, though he would become practised in avoiding the subject among his fellow-philosophers. His monologues gave the impression of unbounded flight, but he would carefully steer them away from the chasm between reason and faith: as he said later of his relations with William Wordsworth, 'in general conversation and in general company I endeavour to find out the opinions common to us – or at least the subjects on which difference of opinion creates no uneasiness'.[20] If the question of faith could be sidestepped, Beddoes' freethinking was in many ways more congenial to Coleridge than the doctrines of the established church. The dissenting community to which Coleridge addressed himself was essentially a coalition between rationalists in the mould of Beddoes, Darwin and his recent mentor William Godwin, and dissenters, Quakers and Unitarians such as Joseph Priestley, who engaged with the same political causes – peace, political reform, the abolition of slavery – but from the standpoint of a devout faith in divine justice. Beddoes, whose social and political networks had always set him alongside Quakers and dissenters, could be as tactful as Coleridge in avoiding divisive doctrinal arguments, and Kant helped them both to square the circle: for Beddoes, he advanced the cause of evidence against revealed truth; for Coleridge, he aligned the same cause with a mission of Christian renewal.

Rodney Place was also a salon for poetry, a passion that Anna shared with the Pantisocrats, and Beddoes returned to dabbling in verse just as he had in the first flush of his friendship with Darwin in 1792. The young men were all preoccupied with composing declamatory verse dramas – Southey had already sold his *Joan of Arc* to Joseph Cottle and was planning another political epic, *Wat Tyler*, while he, Coleridge and Lovell were all three collaborating on their *Fall of Robespierre*. But now, all wed or engaged to the Fricker sisters, they were also playing with shorter lyric pieces, and it is in this category that the surviving fragments of Beddoes' attempts belong. The precise dates of composition are unknown, but within them are flickers of the form that would make Coleridge's name, and that may reflect an early phase of its gestation. They include, characteristically, some parodies, such as 'Receipt to Make a Good Legendary Tale', a satirical inventory of the gothic cliché that permeated the new German *Sturm und Drang*:

> In woods immersed beside a lake
> A very ancient castle take
> And plant therein a damsel fair
> With eyes sky-blue and auburn hair.[21]

To these are added chill nights, a 'pale-faced moon', heaving sighs and errant knights: the medieval and mythic palette that Coleridge would soon, with poems such as 'Christabel', turn to haunting metaphysical narratives of possession and evil. But Beddoes also produced delicate miniatures, such as 'Domicilary Verses', whose opening –

> Invitingly yon single-storied cot
> Peeps o'er the frosted heath[22]

– introduces a rustic cottage at midwinter, within which unfolds a scene of domestic contentment that would find its consummate expression in Coleridge's 'Frost at Midnight'.

*   *   *

By the spring of 1795, with their *Fall of Robespierre* still unsold, the Pantisocrats had decided to raise the funds for their utopian adventure by delivering a series of public lectures in Bristol. Joseph Cottle, struck by Coleridge's aura of 'commanding genius', had offered him, together with Southey, an advance of 30 guineas on their poetry almost at first meeting, but Coleridge had started borrowing extra sums immediately and Cottle, at least by his later account, had already realised that the banks of the Susquehannah were destined to remain over the horizon, since even modest shared lodgings in Bristol seemed permanently beyond the poets' means. The model for their new venture was John Thelwall, the founder member of the London Corresponding Society, who had been acquitted with Hardy at the previous autumn's treason trial and who had since demonstrated that, in London at least, lectures on political subjects to working men could raise consciousness and earn a living at the same time. It remained to be seen whether the same trick would work in Bristol, and how its citizens would respond to the new breed of dissenters.

Admission to the lectures was set at a shilling, the same amount that Thelwall charged in London; and for some at least, they were well worth the price. In a series of inns and rooms over the Corn Market, Coleridge orated with extraordinary fluency against the slave trade and the war. He accused Pitt of manufacturing conspiracies for greedy and corrupt ends; he declared the deported Edinburgh reformers Muir and Palmer martyrs to the cause of liberty and justice, and insisted that all who called themselves Christians should speak out for the abolition of slavery and for peace. One Bristolian recalled twenty-five years later that Coleridge had 'appeared like a comet or meteor in our horizon'; he also recalled that, despite the repeated denials of the lecturer's older self, the views expressed had been 'positively and decidedly democratic'.[23] The Bristol *Observer* proclaimed that Coleridge 'spoke in public what none had in this city the courage to do before, he told men that they had rights' – though the paper also noted, in less stirring terms, that he 'would do well to appear with cleaner stockings in public'.[24]

But Coleridge's lectures also showed that there were many in Bristol for whom speeches on political reform were a provocation that demanded a firm response. The supporters of Church and King were frequently more vocal than his own audience, and so 'furious and determined' in their reaction that he began to fear 'that the good I do is not proportionate to the evil I occasion'. He received meagre rations of the adulation of which he had dreamed, but plenty of 'mobs and mayors, blockheads and brickbats, placards and press gangs' thronging the entrances of the pubs and assembly rooms where he lectured, 'scarcely restrained from attacking the house in which the "damn'd Jacobin was jawing away" . . . two or three uncouth and unbrained automata have threatened my life'.[25] He stood up bravely to hecklers, responding to one that 'I am not at all surprised that, when the red hot prejudices of aristocrats are suddenly plunged into the cool waters of reason, they go off with a hiss!'[26] But it was a reminder that Bristol was a city whose vested interests were still dominated by trade and slavery, and that those speaking in public against them could expect to be met with force.

In the meantime, Beddoes' tireless promotion of the Pneumatic Institute was beginning to pay dividends. 'I seem to be getting into fame', he wrote to Giddy.[27] He had a steady stream of patients, and 'innumerable applications from distant parts': he was reaching Erasmus Darwin's level of reputation, where the class of patients with the means to pay for the best possible care were reading his books, absorbing his theories and requesting his services from all over the country. Darwin had warned him often enough that this level of practice was strenuous and time-consuming, and needed to be billed correspondingly: it was this, apart from anything else, that enabled him to continue to treat the poor for free. Beddoes, however, still had scruples about squeezing patients' purses, even those who could afford it. 'I get money enough', he told Giddy laconically;[28] he was more strongly motivated by the recognition that 'the pneumatic practice is spreading' as a result of his raised profile. By early summer, too, his heavily pruned and annotated edition of John Brown's *Elements of Medicine* was in the bookshops. It was, he felt, disappointingly light on Brown's biography – the personal papers and memoirs he had been promised never materialised – but it sold well, and would become the standard text of Brunonian medicine.

For the moment, the storms of politics and war were leaving Beddoes in peace – though, he feared, only temporarily. 'The world seems very quiet at present', he commented. 'I am looking first to one place and then to another for new mischief to break out.' Ireland might erupt at any time: with the British government so hated and her army so overstretched, who knew how extensive the Defender networks were becoming, and how advanced their plots for a French invasion? The Mediterranean, though, seemed equally plausible as 'the scene of the Devil's first operations; for you must know', he informed Giddy, 'that I am acquainted with a clergyman who says that the Almighty has relinquished this globe of ours to the Devil as a plaything. I really think the hypothesis deserves some attention.'[29]

Mischief was indeed on the point of breaking out, and a good deal closer to home. The combination of disrupted Atlantic shipping and Pitt's war taxes were hitting Bristol hard, and the poor hardest. Many working families could no longer afford the price of bread; malnutrition and even starvation were spreading among their children; when night fell, crowds took to the streets to listen to angry ballad-singers and gather menacingly outside the Corporation and traders' offices. Among the wealthy, some responded by cancelling banquets and other luxuries and raising subscriptions for poor relief; some, too, began to buy the grain and flour cargoes on the docks and pass them on to bakers at cost. But others continued to sell these cargoes to middlemen and speculators for the highest price, which often meant shipping them straight out of town. The Corporation took no action beyond keeping the discontent in check by arresting the most conspicuous troublemakers and holding the out-of-town militias on standby.

It was on the evening of 6 June that the tensions finally spilled over. A crowd had assembled outside a butcher's shop where large joints of meat were on display in the window; as darkness fell, some began throwing stones, smashing the windows, and others grabbed the meat and carried it off. The chief magistrate arrived at the scene almost immediately, bringing with him a militia who struggled to disperse the angry crowd, eventually seizing four men they identified as ringleaders and carting them off to jail. As each party blamed the other, the stories that emerged revealed a system stretched beyond breaking point. Twenty-one butchers had that very day signed a petition to the mayor claiming that their high prices were due to 'jobbers' and middlemen who bought all the meat from the fairs, markets and farms at profiteering prices, and then inflated them further to sell them on to the butchers, who were forced to hike them further still to afford the rent demanded by 'your Worship for our shambles of stands'.[30] The crowd, in turn, alleged that many of the butchers had been using fraudulent weights and scales to undersell to their customers.

Coming so soon after the turnpike riots at the bridge, this was another alarming signal for the citizens of Bristol that they were moving towards the kind of chaos to which they had long believed their prosperous and contented city was immune. Food riots were common enough in some parts of the country, but it was over forty years since they had had one here. As a port city, Bristol had huge advantages in times of scarcity: it was a straightforward matter for the Corporation to requisition cargo at the docks, purchase it by compulsory order, distribute it where it was needed and order more. This accrued costs to the city exchequer, but was far cheaper than dealing with the damage and loss of trade that followed a riot. But despite – or perhaps because of – the violence at Bristol Bridge, the Corporation had failed to intervene on this occasion, and had allowed the poor to starve. Often the events described as 'food riots' in eighteenth-century Britain were relatively orderly, even formalised affairs, where hungry crowds gathered, took the supplies that they needed, and paid for them politely but firmly at pre-scarcity prices. But the Bristol crowd's reaction on the night of 6 June pointed towards a final breakdown of trust between the city's authorities and its population, and indicated that, unless this trust could be repaired, violence would from now on be the first language of Bristol politics.

Bristol was not suffering alone. At the end of June a resurgent London Corresponding Society, with a flush of new members recruited by the collapse of the treason trials and the unpopularity of the war, held a public meeting in St George's Fields on the outskirts of the city, attended, according to some estimates, by a crowd of 100,000. Delegate after delegate from across the country spoke of spiralling food prices, scarcity and starvation caused by the 'cruel and unnecessary war', the only remedy for which was to 'acknowledge the brave French republic and obtain a speedy and lasting peace'.[31] The government had predicted violence at the meeting, but it had remained orderly: handbills had been distributed warning the crowd to keep the peace, and motions supporting king and country solemnly proposed and passed. Nevertheless, the addresses delivered to the home secretary by its delegates were not acknowledged, and the Home Office began drawing up a plan entitled 'In Case of Invasion' which stressed that 'every attempt at sedition or disturbance' was henceforth to be 'repressed with a high hand'.[32]

On 30 July the *Bristol Gazette* included an insert, written by its editorial committee and designed to be pinned to the walls of homes and workplaces. It was headed 'In the Time of Dearth, 1795' and consisted of two sets of numbered rules. The first, 'Rules for the Rich', began as follows:

1. Abolish gravy, soups and second courses.
2. Buy no starch when wheat is dear.
3. Destroy all useless dogs.

It continued with more home-economy hints, an encouragement to make broths and rice puddings for the poor, and warnings to 'go to church yourselves and take care that your servants go to church constantly', and to give nothing to poor people 'who are idle or riotous or keep useless dogs'. The 'Rules for the Poor' were to: 'keep steady to your work', 'avoid bad company', 'go constantly to church', spend any money on staple foods and none on alcohol, learn how to make soups from leftovers, 'be civil to your superiors', and 'be quiet and contented and never steal or swear or you will never thrive'. It was a list that spoke to an imminent social breakdown that the authorities could no longer head off by coercion: behind it lay the clear but unspoken message that if these rules were not followed, the next series of demands would come not from the rich but from the poor, and not in the form of domestic hints in the local newspaper.

Grain prices inched up through July and August, an annual pattern in advance of harvest; but the weather, as it had been the previous summer, was disturbed, drought alternating with flood, and by September it was clear that shortages would bite by early winter. In October, with Parliament due to resume, large mobs began assembling in the centre of London, milling around Westminster and the seats of government. William Pitt, in his Downing Street residence, was surrounded by an angry crowd shouting 'No war, no Pitt, cheap bread', and pelting him with stones that smashed his windows. On 29 October, George III's carriage was intercepted en route to the state opening of Parliament: its way was blocked by a mob, and in the mêlée there were reports of a gun being fired at the king, and a ruffian in a green coat attempting to drag him from his carriage.

Pitt was not blind to the case for peace, but he feared the backlash that would follow a climbdown: capitulating would force his resignation and bring calls for political reform, which anger and hunger would whip up into violent protest and mob rule. Privately he told colleagues that 'my head would be off in six months were I to resign'.[33] His public response came in Parliament where, in early November, he rushed to propose two bills to deal with the crisis in public order. The first, the Treasonable Practices Bill, extended the definition of treason by removing the need for it to express itself as an 'overt act': from now on, mere writing or even private speech could be held to be treasonous. The second, a clear response to the meeting in St George's Fields, was the Seditious Meetings Bill, under which all unauthorised public gatherings of fifty people or more – the quorum for Corresponding Society meetings – would be assumed to be treasonous, the burden of proof shifting from the Crown to the accused. The two bills – quickly christened the 'Gagging Bills' by Charles James Fox and the opposition – were a calculated statement to the partisans of political reform: not only were no new rights to be granted, but the old ones were to be curtailed.

It was a move that deepened the rift between Britain's two political cultures and also, with its overt attack on the national constitution, drew many previous loyalists onto the reformers' side. For a century, the British had delighted in taunting the absolute monarchies of Europe with their constitutional rights to free speech, a free press and free public gatherings; now, all were being suppressed. Across the country, clubs and associations sprang up to exercise the rights that were to be withdrawn in a matter of weeks. Davies Giddy, promptly nominated deputy-lieutenant of Cornwall and handed emergency powers to deal with the crisis, was typical of many who were thoroughly uncomfortable with the new dispensation they were expected to enforce. Anyone could exercise their constitutional right to free speech in private until November; but by the end of the year the same words could see them jailed, transported, even executed. If Giddy attempted to protect the victims of the new laws, he would become a popular hero in ways that his fellow-officers and magistrates would delight in turning against him.

For Beddoes, it was a defining moment. For years, his anxieties about the abuse of power and patriotism had ostracised him from the mainstream of society and tarred him with the brush of Jacobinism; now, thousands were flocking to stand up and be counted for the political principles that he had defended for so long. It was clear why so many were suddenly committing themselves to the cause: this would be the last chance, as far as anyone knew, to hold a public meeting or express a dissenting political opinion in Britain. With the support of the duke of Portland's Whigs, Pitt's administration commanded an unassailable majority in Parliament: the bills would be law before Christmas.

It was a cruel irony that this moment should arrive just as the Pneumatic Institute had built up such an impressive head of steam; but it was a moment that might never return. This was the time for action, but *what* action was by no means clear. The attempt to involve the mayor and the Guildhall in the peace protests earlier in the year had shown that an official approach was unlikely to provide a public platform for Bristol's dissenting citizens. Yet it would be reckless in the extreme to hold an unsanctioned public rally in a town that was on the brink of insurrection. However just the causes, Beddoes still held a Burkean terror of the mob: after the June food riots he wrote unequivocally to Giddy that 'a riot must be suppressed: that is no less clear than that a conflagration must be stopped'.[34] The Corporation and the Bristol elite might be undemocratic and incompetent beyond reform, but no good would come of inciting the hungry poor against them. The correct forum must somehow be engineered to allow Bristol's reformers to break a lifetime's habit of discretion and concealment and make a public stand.

* * *

At noon on 17 November, the Guildhall had been booked for a 'Meeting of the Citizens of Bristol', which was tabled to propose a Congratulatory Address to His Majesty George III 'on his late providential escape from the attack and insult offered to his person, and to show their utmost abhorrence of such proceedings'.[35] From eleven o'clock onwards, a thin crowd began to file in, and at the appointed hour James Harvey, the mayor, arrived to chair the meeting. But as soon as he had taken his seat, the hall suddenly filled towards capacity with 'a great concourse of most respectable tradesmen', many of them clutching copies of a handbill that had been printed overnight and circulated from early in the morning. A local merchant named Coates asked if he might speak a few words and, as it was an open meeting, the mayor agreed. Coates then proposed that the congratulations to his Majesty should be passed with an amendment, which he read from the handbill: 'to beseech his Majesty to restore the blessings of peace to his faithful people'.

John Rose, a local printer, asked the mayor if he might second the amendment with a brief speech. He concurred heartily with the address, but added, 'I am fearful, Mr. Mayor, that our gracious sovereign knows not the voice of his subjects respecting the present ruinous, disastrous, and calamitous war', this voice being concealed from him by 'wicked, designing and corrupt men'. Rose, a known reformer who had published a pamphlet disputing the official version of the Bristol Bridge riots and challenging the trustees to make their accounts public, was echoed by a stir of voices from the majority of the audience: those who had arrived at the last minute clutching the handbills.

It was now clear to all that the meeting had been hijacked, and that Bristol would, after all, have its debate on war, peace and the Gagging Bills through official channels. The alderman protested on the grounds that it was 'presuming on the power of the executive government to offer sentiments on the propriety of the war', but he was shouted down by voices asserting their constitutional right to express their sentiments to their monarch. The hubbub eventually quietened enough for the sheriff to insist that the government was indeed exerting itself for peace, and that those who wished to force it from its course by violent protest were the type of fanatics who, like the Jesuits, believed that 'the only way to procure a peace was by the sword'.[36] At this the house erupted in groaning and hissing, clapping and stamping, a deafening protest amid which no one could be heard. Eventually the chaos ebbed, to make way for a voice that had been calling 'Mr Mayor! Mr Mayor!' for a considerable time, and that emanated from a wide-eyed young man in a dark, brass-buttoned greatcoat.

Coleridge had had an eventful summer, and was now set on a new course. Pantisocracy was no more: it had been too pure in its ideology and too naïve in its practical foundations to survive the tensions between its founders. Southey, impatient to bring it down from the clouds and make it a reality, had sought to scale it back to a rural commune in Wales; for Coleridge, this had been death by a thousand cuts, a cheapening of their vision that showed Southey had never been worthy of it. In September Coleridge had met William Wordsworth, whom he had swiftly cast to replace Southey as the solid counterweight to his mercurial wanderings; in October he had married Sara Fricker, and was now on the point of leaving Bristol for domestic isolation in the village of Clevedon, which overlooked the Bristol Channel. During the week of the Guildhall meeting he had broken decisively with Southey ('You are lost to me, because you are lost to virtue . . . this will probably be the last time I shall have occasion to address you');[37] now, addressing a crowd many times larger than at his lectures, he was presenting his new incarnation in public for the first time.

Coleridge's speech to the crowd was long remembered as 'the most elegant, the most pathetic, and the most sublime address, that was ever heard' at the Guildhall.[38] The assault on the king, he agreed, was a great evil; but it was also the consequence of a great evil, the suffering of the people under 'the present cruel, sanguinary and calamitous war'. The poor were unrepresented at the current gathering, but it was on them that the brunt of war fell. A gentleman might lose some of his property, but he would also be left with some; 'a penny taken from the pocket of a poor man' might be his all. Coleridge continued until 'authoritatively stopped', at which point the mayor insisted he would quit the chair unless the noise abated and the peace amendment was dropped. There were many who wished to speak – Robert Lovell had to be 'personally prevented' from taking the floor – but eventually Coleridge was granted space for a few final words. The paradox, he concluded, was that 'the very means which government had taken to prevent Jacobinical principles, were calculated to produce them'. The bulwark against revolution in Britain had always been her constitutional freedoms; if these were obliterated, she would become like the *ancien régime* of France, and with the same future in prospect.

The meeting descended again into chaos, with those gathered shouting at the mayor to ask when they would be allowed to use the Guildhall to discuss their motion. No answer was received; the mayor left the chair, and the meeting broke up.

*　*　*

The following morning, a new pamphlet appeared in the bookshops of Joseph Cottle and John Rose, and on the stalls around the Corn Exchange, where it did brisk business. It was entitled *A Word in Defence of the Bill of Rights against the*

Gagging Bills, and its author was Thomas Beddoes M.D. He had learned much since dashing off his handbill after the September Massacres: this was a much more measured tract, thorough in its argument, pitched squarely at moderate opinion and argued patiently from first principles. 'Reason and speech', Beddoes began, are 'the two great faculties by which the Almighty has distinguished man from the brute creation',[39] and it was no coincidence that it was the right to use these faculties – a right by which 'the indiviudal inhabitants of Great Britain have been distinguished for a century past' – that had made the British constitution the envy of the civilised world. It was no novelty or revolutionary claim, but an inheritance 'derived from our ancestors, by whom it was dearly bought, as by us it has been securely enjoyed'. Edmund Burke might argue that avenging 'the beautiful and high-born Queen of France' is an obligation that demands the suspension of these rights, but it remains to be seen whether the British people are equally 'astonished at the magnitude of this outrage', or whether they are more persuaded that 'a nation which slumbers over its rights, will be fortunate if it awake not in fetters'.[40]

For it is to fetters that the Gagging Bills must lead. 'To be debarred from assembling' is effectively 'to be debarred from ever carrying a petition adverse to the will of the minister'; this, in turn, is an invitation to corruption, since 'the more incapable, designing or dangerous the minister, the more certainly will he guard his follies or his villanies' with 'a half usurped and half delegated despotism'. To surrender the right to interrogate the government is ultimately, not merely for the poor and voiceless but even more for the propertied classes, 'deliberate political sucide'.[41] It will give *carte blanche* to every local magistrate to act as 'the tools of an unprincipled ministry', using 'that law which is to give them despotic power over you when assembled'. Already the authorities are attempting to prevent these arguments from being heard through official channels; if they persist, Beddoes concludes, 'Citizens! let not petty difficulties prevent the exercise of your rights! Assemble, if you can find no other place, in Queen-Square.'[42] Summoning an angry crowd onto the streets was not a call that Beddoes made lightly; but for all his fear and hatred of the mob, this was a moment when everything must be risked if liberty was to be preserved.

The mayor was now under pressure from all sides: the crowds at the Guildhall meeting, Beddoes' pamphlet, the groundswell of civil dissent that they articulated and the undercurrent of street violence that formed their ominous backdrop. He decided to soften his stance, and offered the Guildhall for an official public meeting at noon on 20 November to discuss 'the propriety of petitioning Parliament against certain bills now pending in the House of

Commons, by which bills it is conceived that the Bill of Rights will be invaded'.[43] From early in the morning it was clear that this time both sides would be prepared for the fight. Fierce storms had swept across the country for the previous three days, disrupting moorings and strewing wrecked ships across the docks; shops and houses were still battened down as the county militia paraded through the streets, while 'vast numbers of handbills were distributed to prevent the friends of peace from attending'. At the door of the Guildhall the delegates were obliged to file past a picket of loyalists beating drums to summon the supporters of Church and King.

Inside, the sound of the drums became muffled and 'a pin might almost literally have been heard to drop in any part of the Hall' as the proceedings got under way. They were chaired by the doctor Edward Long Fox, a Cornish Quaker who had patiently worked his way into the closed society of Bristol's professional elite. In 1796 he had become the first incomer to break the local monopoly and win an appointment at the Bristol Royal Infirmary; recently he had set up the city's first private asylum for lunatics, where patients were treated humanely according to the precepts of the Society of Friends. He had already made himself conspicuous as a reformer, particularly in the aftermath of the bridge riots when he had led the campaign for a public meeting at which the bridge trustees might answer their critics, an offer the trustees had refused. Now, he echoed Beddoes in his call for action to preserve the 'true, ancient and indubitable rights' of British citizens to the 'peaceable assembling of the people together'. He insisted that the meeting should be conducted with 'order and decorum', and opened the floor.

The first resolution, proposed by a banker named John Savery, requested that the meeting should express 'the utmost abhorrence' at 'the late outrage and insult on his Majesty'; Robert Lovell relayed this sentiment at loud volume to the back rows of the hall, and it was passed by a unanimous show of hands. There would thus be no grounds for anyone, either within or outside the meeting, to doubt its patriotic intent. Savery then presented a petition against the Gagging Bills, which was seconded by Beddoes and also adopted without a dissenting voice. But the question that now arose was to whom the petition should be presented. Coleridge proposed that it should be handed to Charles James Fox and Richard Sheridan, the leaders of the Whig rump who had 'ably and unequivocally stood forward to oppose the present bills'. Beddoes, however, argued that it would have more weight if it were presented by 'our proper representatives', the two Members of Parliament for Bristol, directly to Pitt's administration, rather than to the minority opposition. Coleridge acceded and, though there was 'some hissing' at the suggestion of involving Bristol's MPs, the resolution was adopted once it was made clear 'that doing so would by no means pledge themselves to support them at a future election'.[44] Votes of thanks were passed, and all the attendees silently and solemnly signed the petition and vacated the hall in a pointed demonstration that Britain's constitution might at this point be safer in the hands of its people than its government.

Just as the first meeting had been followed by Beddoes' pamphlet, so the second was chased up with a squib from the opposing faction, an anonymous pamphlet entitled *A Letter to Edward Long Fox M.D.*, but with Beddoes and Coleridge equally in its sights. It accused Fox of stirring up 'the citizens of Bristol in a way calculated to mislead their judgement, inflame their passions and excite their resentment',[45] without first asking himself whether he had 'sufficient influence, authority or abilities to stay the career of a thoughtless populace'. It had been a highly irresponsible act by one who must know 'how much more of passion than of reason falls to the lot of the lower ranks in every society'; but, despite Fox's well-known and sinister interests in animal magnetism and French politics, the citizens of Bristol were not to be mesmerised into revolution. The anonymous author was confident that the local populace had more 'spirit and prudence' than to 'suffer a few factious aliens to scatter among them the seeds of discord and sedition'.[46] Fox, Beddoes and Coleridge, he reminded his readers, were all outsiders with their own agendas; if their agitation reduced Bristol to a smoking ruin, they could simply move on to the next city.

The pamphlet war rumbled on with the appearance of an anonymous response to the *Letter to Fox* signed with the suggestive initials of 'S.C.T.' and now generally attributed to Coleridge; but the opportunity to turn the tide was fading. On 18 December the Gagging Bills became law. The opposition to them in Bristol had been mirrored across the nation in campaigns, protests and public meetings that had produced nearly a hundred petitions and generated a momentum that showed no signs of winding down. In Manchester the Two Acts, as they were now known, were challenged by the formation of a Thinking Club, which three hundred members attended but at which not one word, treasonable or otherwise, was spoken.

At the Whig Club in London on 20 December, Thomas Erskine, who had so heroically defended Thomas Hardy in the treason trials, urged that every friend of freedom should act as 'a species of watchman' to protect Britain's constitutional rights, now that the law and the government had abandoned the post.[47] During a meeting in the Rummer Tavern on Bristol High Street a few days later Erskine's call for a watchman was repeated, and several of the company urged

Coleridge to gather his opinions, lectures and poems, together with those of his literary circle, into a periodical that might serve as a guardian of liberty's flame.

*   *   *

On Christmas Day, Beddoes wrote to James Watt with, as was becoming his habit, his unvarnished thoughts, relieved of his public necessity for advocacy and projection. 'I know very well', he began, 'that my politics have been very injurious to the airs'. Yet he still did not see how he could have acted otherwise. It was not as if pneumatic medicine could truly be separated from politics: 'as every stroke aimed at liberty equally threatens science, morals and humanity, it requires great self-denial to look on patiently and silently when such great interests are at stake'. Now, however, with the public mood turning, he was becoming more sanguine than ever about his project. The causes with which he had become associated were standing 'on more popular ground than ever before'; might not the Pneumatic Institute now reap the benefits of his courage in associating himself with them even when they were unpopular?[48]

This was, however, a justification that tacitly acknowledged there were many, Watt included, who had seen no need to join the political fray in defence of science, and some who had found Beddoes' actions gratuitous and regrettable: Joseph Black, for example, had told Watt in July that he was 'sorry to see that Beddoes is so absurd and wrong-headed as to set himself up as a statesman and attack Mr. Pitt'.[49] Beddoes' justification was not an apology; nor was it a promise to desist. His *Word in Defence of the Bill of Rights* had been the best seller in the local pamphlet wars, and he had already followed it up with a sequel, entitled *Where Would Be the Harm in a Speedy Peace?*, in which he had rallied the 'Inhabitants of Bristol! You, who lately stood forth the asserters of liberty!',[50] to join together to oppose a war that was condemning the nation to bloodshed and debt for a generation. As the year turned and 1796 blew in on yet more prodigious storms and gales, the grain shortages bit harder and, surrounded by malnutrition and disease, Beddoes found his political and medical instincts winding ever more tightly together.

By 12 January he had another pamphlet on the streets of both Bristol and London. *A Letter to the Rt. Hon. William Pitt*, subtitled *On the Means of Relieving the Present Scarcity and Preventing the Diseases that Arise from Meagre Food*, combined his polemic against the war with medical advice for the emergency that it had generated. It began with an extended harangue, laying the blame for food shortages squarely at Pitt's door: 'Did it never, sir, occur to you that unproductive years were to be guarded against? or did this contingency escape you at the moment you entered upon measures, which unavoidably enhanced the national demand for produce?'[51] The French had bought up grain exports from America and the Mediterranean, while Pitt had done nothing ('what miscalculation on your part!');[52] cold, 'the great ally to hunger in the business of destruction',[53] was upon the nation, and yet the rich were still wasting its resources on luxuries, and refusing to take responsibility for the hardships that surrounded them on all sides. If Pitt, or indeed 'Mr. Burke or even Miss Hannah More', would take the trouble to 'write an exhortation to the purpose, the manufacture of fringe and varnished tables might give way to worsted stockings and flannel jackets', and other necessities that 'would preserve thousands from death, and redeem tens of thousands from suffering'.[54]

If Pitt would not listen to Beddoes the politician, Beddoes the physician still had practical schemes to offer, presented in the plain speech of his *Guide to Self-Preservation*. When starvation threatens, it is to our meat consumption that we must look first, as 'every ounce of beef contains the quintessence of many tons of grass, hay and turnips'.[55] All these crops should be switched to potatoes, and every last shred of meat should be recovered by 'broth-machines', pressure cookers able to extract marrow and nutritive juices from scraps that were currently just discarded. Beddoes presents diagrams of his machines, tables of the pressure they exert – 'six atmospheres, or 84lb the square inch' – and recipes for making 200 quarts of 'very good wholesome palatable soup' from half-a-crown's worth of shinbones, potatoes and carrots.[56] Much could also be achieved with regulations and taxes on alcohol, where barley is 'as much wasted in the preparation of strong beer, as corn in the distillation of spirits'.[57] It may even be necessary to consider more general use of opium which is, according to Dr Darwin, 'to a certain degree capable of supplying the place of food', and thereby to preserve life through periods of acute shortage.[58]

A year previously Beddoes had had no time to engage with politics, but now it had forced its way to the forefront of his agenda. The moments when the public were prepared to engage *en masse* with these wider questions were, he knew, the exception rather than the rule, and there was no knowing how long they would last; but while they did, the opportunity to foster them must be taken. Now, too, he had the ideal London publisher: Joseph Johnson, who had been buying Beddoes' Bristol pamplets from Joseph Cottle and selling them from his shop in St Paul's Churchyard, decided to print his own edition of the *Letter to Pitt*. Johnson had for years been publishing the work of Joseph Priestley and Tom Paine, but since the treason trials of 1794, during the course of which he had distributed works by the accused including Thomas Hardy, John Thelwall and Horne Tooke, he had been deserted by many of his literary clients. He was now the leading specialist in the busy, if precarious, niche of dissenting works that tested the new strictures of the Two Acts.

Within a few weeks Beddoes had another, and far more substantial, political commentary in the press: a two hundred-page *Essay on the Public Merits of Mr. Pitt*, offering a complete history of the prime minister's career and its effects on Britain. He recalled Pitt's early days, when he had been a leading champion of political reform, no less than 'the Hotspur of innovation'; he reminded the public that it was Pitt who had once warned that, without fair and proper representation of the people, the House of Commons 'would degenerate into a mere engine of tyranny and repression'.[59] But in his pursuit of power all of Pitt's principles had evaporated, and his politics had shrunk to a cynical 'commercial project': the dispensing of bribes, offices, titles and contracts to forge a coalition of vested interests impervious to the will of the people. Now, as a war leader, he had unprecedented levers of power at his disposal, but he had used them ruthlessly to raise taxes – including, with disastrous effects on public health, a window tax – that had oppressed the population and transformed him into a fully fledged tyrant.

Beddoes' exhaustive catalogue of Pitt's errors concludes with a 'melancholy conjecture' of where Britain might be now, had the prime minister kept the promises of his youth and 'been the author of benefits to mankind'.[60] We would have had 'gratification of that desire to redress the wrongs of Africa', the support of which by 'the great majority among us' is now drowned out by the slavers and free-traders with whom Pitt, Dundas and their colleagues have forced the nation into a relationship of dependency. Had Britain acted with the principled humanity that her citizens demanded, our example would have 'disarmed the internal enemies of France', which would now be governed by the moderates of 1789 rather than the terrorists of 1792 or the military despots currently tightening their grip. Britain and France might then have forged ahead together as constitutional monarchies, and by their example be liberating the rest of Europe from despotism rather than allying with its worst tyrants. Britain herself would have been transformed by the innovations that have instead been suppressed, and would be winning 'applause instead of condemnation from mankind'[61] for her enlightened influence on world affairs. It was an elegy for a lost future, one that had burned so brightly in Beddoes' mind for so long, but was now no more than a sigh of loss.

* * *

Like Beddoes, Coleridge was determined to maintain the momentum of the public opposition to the war and the political clampdown. Over New Year 1796 he decided to produce a periodical that would defy the Two Acts by holding the government to account and forcing, if necessary, a public test of the new laws that might be as damaging to the government as the treason trials had been. In reference to Thomas Erskine's exhortation, though also for its resonances with the prophets Ezekiel and Isaiah, it would be called *The Watchman*. While Beddoes was writing his chronicle of Pitt, Coleridge was touring the country, selling subscriptions to his new venture in Birmingham and Manchester, Derby and Nottingham. Much of his itinerary was arranged by the same network through which Beddoes had been fundraising for the Pneumatic Institute and he even had the pleasure of an overnight stop with 'the most inventive philosophical man' in Europe,[62] Erasmus Darwin, whom he admired greatly as a poet, though he was infuriated by his amused tweaking of religion, and particularly his mischievous characterisation of Unitarianism as 'a feather bed to catch a falling Christian'.[63] Coleridge made a huge impression among the Unitarian communities on his travels, delivering magnificent sermons 'precociously peppered with politics'[64] to audiences of sometimes over a thousand, and joining religious support to that of Beddoes' and Darwin's philosophical circles.

The Watchman was launched on his return to Bristol in March, with a handsome roll of subscriptions and the ringing banner motto 'That All Shall Know the Truth, and That the Truth Shall Make Us Free'.[65] It was scheduled to appear every eight days to avoid the stamp tax that fell on weekly publications, and its content was an ambitious confection of political essays, poetry, book reviews, foreign and domestic news round-ups, digests of reportage from London and Paris, and reprints of handbills from local civic protests. From the beginning Coleridge was heavily dependent on other, mostly anonymous contributors, of whom Beddoes was among the most prolific: his identifiable contributions include various reviews and essays on scientific and medical topics, and an essay on Kant's new epistemology. Beddoes' books were, in turn, generously endorsed: in the third issue, his *Letter to Pitt* was summarised at length, with the praise that 'to announce a work from the pen of Dr. Beddoes is to inform the benevolent in every city and parish that they are appointed agents to some new and practicable scheme for increasing the comforts or alleviating the miseries of their fellow-creatures'.

It was not long, however, before the ambitions and risks of *The Watchman* began to exert an unbearable strain on its editor. The subscriptions tour had been bad enough: while staying with Darwin at Lichfield, Coleridge had confided that 'I verily believe no fellow's idea-pot ever bubbled up so vehemently with fears doubts and difficulties than mine does at present'.[66] By the time the first issue was out and the reality of producing thirty-two pages a week had begun to sink in, apprehensions had been replaced with terror. 'I have been tottering on the edge of madness', he wrote the day after the first issue appeared – adding, in his first written reference to a lifelong theme, 'I have been obliged to take laudanum almost every night'.[67]

Joseph Cottle, who had sold 250 subscriptions in Bristol, around half the overall total, remarked with hindsight that 'of all men, Mr. Coleridge was the least qualified to display periodical industry', and that consequently 'a feeling of disappointment, early and pretty generally, prevailed among the subscribers'.[68] Coleridge recognised soon enough that he had set himself a virtually impossible task, and persisted only out of a general sense of duty to the cause, and in the teeth of overwork, debt, terrible anxiety about the consequences of supporting a family, and constant fear of prosecution. As the issues rolled on and the financial reserves to pay contributors dwindled, the content began to thin: becoming lighter on political essays and original poems, and heavier on scrappy recyclings of other newspapers, digests and, indeed, itself. By May, and the tenth issue, despite a fresh whip-round from Cottle, debts could no longer be met and the project folded with Coleridge's despairing sign-off, 'O! Watchman! Thou hast watched in vain!'

It was a harsh verdict, tinged with both self-pity and self-blame; yet in many ways it was not the journal whose promise was unfulfilled, but the moment on which it had sought to capitalise. The Watchman had perhaps not been as boldly confrontational as many had hoped: the war, and even the Two Acts, had featured in it less and less as it progressed. Its pious tone, with its assaults on government, war and slavery wrapped sonorously in the language of Christian duty and moral rectitude, had grated on the secular wing of its readership; for some the poetry alone had been worth the cover price, while for others the poems had been digressions of baffling obscurity among the grist of reportage. Yet in many respects it had stuck bravely to its mission, mounting several savage personal attacks on Pitt, declaring the Two Acts unconstitutional, supporting the rights of man, looking forward to the overthrow of tyrants, declaring property to be theft from the poor and producing perhaps the most eloquent attack on slavery of its day. Twelve years later Coleridge would quietly recover many of its original elements for the great journal of his mature career, The Friend.

By May of 1796 it was clear to both Beddoes and Coleridge that they were becoming beached by political dissent's outgoing tide. The previous winter's rush to the barricades had ended; as Beddoes had feared, the ardour stoked by the Two Acts had proved to be of limited span, and the sides were no longer as clearly drawn as they had been at the end of the previous year. The case for peace was now widely accepted: despite the intransigence of Grenville, Dundas and other hawks in his administration, Pitt had begun tentative negotiations with the French. Nor could the French any longer be held up as an alternative to British belligerence: since the formation of the new Directory in October, with the young general Napoleon Bonaparte as its rising star, France had instituted censorship and banned political meetings just as Britain had. The reformers' gloomy prophecies were coming to pass: the choice between Britain's repressive monarchy and France's military dictatorship was now an impossible one, and would remain so for the foreseeable future. Faced with this impasse, Coleridge, like so many of his generation, would take a winding detour away from political agitation, towards a quietist retreat from the world and a private regeneration of the soul.

But for Beddoes the choice was a different one: between healing society by the swift and drastic surgery of politics or by the slow and patient improvement of medicine. It was a dilemma that had nagged at him throughout his venture into political engagement. At the opening of his *Essay on the Public Merits of Mr. Pitt*, he recounted a comment that a friend of his had overheard during the agitations of the previous November. 'So Dr. B was at a meeting of this committee last night', a respectable lady had been saying. 'Bless me, suppose he had been suddenly wanted! The poor patient might have perished before he could be called. I wonder what a physician has to do with politics?'[69] It was a concern to which Beddoes had a characteristically waspish response – 'when a physician's attendance is unexpectedly desired, if he be found bearing a hand of whist, or partaking of a turtle, he will incur no reproach' – but it nevertheless raised a genuine dilemma. As a physician, Beddoes' primary duty of care was not to the government, but to the health of its citizens. He had engaged in a battle that, though it had ultimately proved unwinnable, had nevertheless had its successes: he had given Bristol its most eloquent dissenting voice, and his *Essay on the Public Merits of Mr. Pitt* was still 'making its way silently out into the world'.[70] Now, though, it was time to return to breathing life into the Pneumatic Institute.

The momentum he had built up had not gone entirely to waste. From Göttingen, Christoph Girtanner was sending news of more successful pneumatic treatments; there was interest now in Vienna, from where he was receiving urgent technical queries about the apparatus. James Watt was continuing to refine his designs, and had come up with a simplified model which,

Beddoes told Giddy with characteristic optimism, 'will suit a kitchen or parlour common fire'.[71] His educational side projects, with the encouragement of his fellow-enthusiasts Richard Edgeworth and Tom Wedgwood, were also beginning to take shape: he had hitched his ideas about manufacturing 'rational toys' to the treatise of a Bristol mathematician, who included Beddoes' proposal as an appendix to his own scheme for constructing geometrical models. 'All ideas are derived from sense', Beddoes began, in Hartley's familiar terms, and proceeded to reason that children should be plied not with useless poppets and decorative figures, but with 'models, at first of the most simple, and afterwards of more complicated machines': engraving tools, letterpresses, mechanical models and chemistry sets.[72] Edgeworth and his daughter Maria, Anna's older sister from her father's previous marriage who was just starting to turn her talents towards a literary career, pitched in with Beddoes to engage artists and build prototype models, though the project stalled in its search for a manufacturer.

* * *

By summer the constellation that had formed over the previous year, and had exhausted itself through the battles of the winter, began to drift out of focus. The diaspora was presaged by the shocking death of Robert Lovell, aged only twenty-four, who caught a fever in Salisbury, decided to make for home rather than rest, and never recovered from his journey. With the estrangement between Coleridge and Southey, the Pantisocratic fellowship was dissolved and Coleridge moved on, attempting to put both Bristol and politics behind him. Among the many he had met in the course of his political activities was Tom Poole, from the village of Nether Stowey in the Quantock Hills to the south of Bristol; over the spring and summer he had spent much of his time in Poole's rural retreat, and would relocate formally to Stowey by the end of the year.

In many ways Poole was, for Coleridge, a rustic version of Beddoes, and he would take on a similar role as his mentor. Like Beddoes, he came from a family of tanners and, like Thomas Beddoes senior, had built up the family business to become a prosperous landowner. His house, set back from Stowey's main street, had a similar aspect to Rodney Place, with its warm sandstone façade and simple pillared portico, and also included a large library where Coleridge sought sanctuary from the pressures of work, debt and marriage. Poole's history of political engagement, too, had run in close parallel with Beddoes'. He had been a conspicuous reformer and enthusiast for the French Revolution in a conservative rural society; as his cousin Charlotte recorded after one of his visits, he 'is never happy till he has introduced politics, and as usual he disgusted us'.[73] Since

1793 he had also been leading local campaigns for the abolition of slavery and against the war with France, and had supported ethical campaigns to boycott sugar and substitute it with honey. He had been charmed and inspired by the bubbling utopianism of the Pantisocrats; he had come to share their conviction that there was nothing left for Europe but war and despotism, and had regretted that he was too irrevocably tied to the land to join them. The Terror and subsequent events in France had plunged him into a long and deep depression in 1794, from which he was emerging just in time to nurse Coleridge through the profound crisis of confidence that followed the failure of *The Watchman*.

If Coleridge was shifting his centre of gravity from Beddoes to Poole, he was not abandoning the passions that Beddoes had awoken in him. 'I am studying German', he wrote to Poole in May 1796, 'and in about six weeks shall be able to read that language with tolerable fluency'. New projects multiplied as ever, now including a translation of the complete works of Schiller – though, as Cottle recalled with exasperation, 'to project, with him, was commonly sufficient . . . while the vividness of the impression lasted, the very completion could scarce have afforded more satisfaction than the vague design'.[74] Yet the seeds that Beddoes had planted in Coleridge would in time bear fruit, and he would return to fertilise Beddoes' own project in unexpected ways.

The summer also brought the departure of Tom Wedgwood, sufficiently recovered under Beddoes' ministrations to undertake a convalescence tour to Germany accompanied by his former chemistry tutor John Leslie. They travelled together through the Swiss Alps, an experience that would have seemed impossibly beyond his reach a year earlier but now stimulated one of the healthiest and happiest interludes of his life. He walked much of the route, at one point covering 200 miles in a fortnight, and sent his brother John almost daily letters crowded with his impressions of the bracing air and natural majesty of the mountains.

The summer of 1796 also marked five years since Beddoes and Giddy had danced in Cornwall with cockades in their hats; it was hard to believe that the world had changed so far and so fast. The anniversary set both on a train of reflection, and each marked the event in his own way. Beddoes memorialised Bastille Day with a poem entitled 'Verses on a Cornish Lady', which attempted to breathe life back into those last golden moments before the world had turned, and celebrate them with the wish:

May Freedom's garland still as now
O'ershadow Beauty's burnished brow.[75]

Giddy, meanwhile, had little to celebrate. The date was a reminder that he had spent five bruising years conscripted into a conflict he had come to loathe. He was now under orders to fortify Cornwall against French invasion, calling up volunteers and converting St Michael's Mount, the island in the crescent bay that had formed the magical horizon of his childhood, into a cannon outpost. Returning exhausted to Tredrea, he found the family home full of drunken revellers, freemen who had been receiving the hospitality of the incumbent MP, 'gorging themselves with meat' and 'drinking until they can swallow no more' in return for their votes.[76] Over the past five years he had trudged the path of duty, and had been rewarded with success and status; but the manoeuvres, compromises and evasions this had entailed had never ceased to stick in his throat. 'I took down a plan of the Constitution of France', he noted in his diary, that 'was stuck up against the wall of my sitting room by Doctor Beddoes and myself in 91'.[77] For Giddy, the promise of that golden season had soured beyond recovery.

PART 2

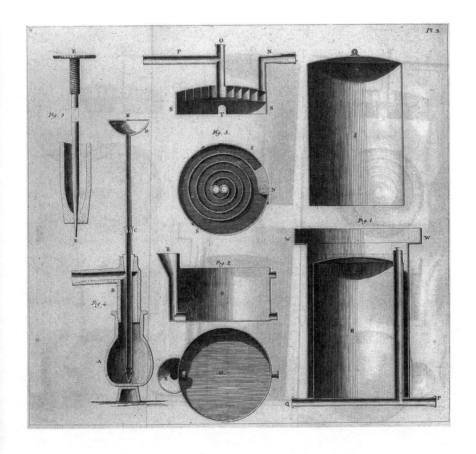

5

The Extraordinary Person from Penzance

As the summer of 1796 drew to a close, Beddoes returned to the Pneumatic Institute with, in the words of his first biographer, 'increased activity, if that were possible'.[1] A thick new version of *Considerations on the Use of Factitious Airs* went to press, including this time a London edition from Joseph Johnson: the previous case studies had been expanded to impressive length with updated notes from medical contributors, and new appendices introduced researches on pneumatic medicine and related chemical studies from around the globe. Among these, the American chemist and fellow-Brunonian Samuel Latham Mitchill had made trials with a new gas, an oxide of nitrogen that he believed to be the component of the atmosphere that bred putrefaction and decay; and from Bombay, Dr Helenus Scott, a surgeon with the East India Company, had found in repeated trials that nitric acid was effective in healing the sores and lesions of syphilis, a 'loathsome disease' that most infirmaries refused to treat and for which private doctors could only offer corrosive and toxic mercury salves. All such exciting preliminary findings were grist to Beddoes the projector, selling as they did the perception and the promise that pneumatic and chemical medicine were on the march – though Beddoes the empiric, freed from the necessity to believe in his own theories, would have admitted that they still required considerable experiment to validate them.

Beddoes was also looking for new premises. Hope Square was an adequate workshop, but far too small for his new plan 'to assemble everything mechanics, chemistry and art in general can supply for the relief of the sick' into a single combined laboratory, outpatients' clinic and residential hospital.[2] But the undertow of war, economic slump and political crisis was still holding back his financial prospects. His subscriptions had slowed after the initial surge of 1795, and the £1,400 or so he currently had at his disposal was well short of the amount needed to launch an Institution whose ambitions had grown faster than its means. 'The distressed state of the country' was thinning the sick trade at Hotwells, and with it both his earnings from private practice and his chances of finding volunteers for 'a design that bids fair to do good and lies within the reach of my own means'.[3] The tentative peace talks between Britain and France had broken down, and the French army's surge into the Italian peninsula fed a fatalistic sense that it was only a matter of time before Britain would fall. 'We shall be invaded, I believe', Beddoes informed Giddy gloomily; the French would aim simultaneously at half a dozen points on the coast between Devon and Northumberland, and 'will undoubtedly effect their landing in 2 or 3 places at least'.[4]

He may also have been, he realised, counting the cost of his political adventures. Although he was perhaps the most eminently qualified and certainly the best-known physician in Bristol, the city's professionals remained reluctant to open their doors to him. His application for membership of the Bristol Philosophical Society should have been a formality, but in November 1796 it was refused on the grounds of his 'opinions in regard to religion'.[5] Erasmus Darwin might have built his towering reputation while maintaining a genteel and discreet atheism, but that was before the turbulence of the 1790s, and even he was now drawing in his horns. For Beddoes, who had taken sides more conspicuously and in more divisive times, it was proving much harder to eradicate the print of his cloven Jacobin hoof.

With all these frustrations and uncertainties, Beddoes' energies had begun, as was their wont, to dissipate into other channels. He was organising a drive to use nitric acid in the treatment of syphilis, which was being pioneered at Plymouth Hospital, where doctors had found it 'certain, expeditious and infinitely milder than mercury'. Beddoes was now coordinating a call for more experiments, and for case histories to assemble into a book.[6] But he was also beginning to expand his pneumatic ambitions further, and to conceive of the Institution as only a part of the vast and underexplored field of preventive medicine into which he had ventured with his educational tracts such as the *Guide to Self-Preservation*. The rich were well aware that the beneficial airs of mountains, coasts and spas could arrest the early stages of conditions such as consumption, the only point at which the disease could be treated with anything approaching effectiveness. The poor, by contrast, were defenceless against not only consumption but other diseases of the chest whose incidence multiplied every year from exposure to toxic air in mines, factories and choking inner cities. Yet if pneumatic therapy could be developed to function as prevention as well as cure, might not those who could never afford a spa have beneficial airs administered to them in their slums and workplaces?

This would represent a sweeping and revolutionary extension of the Institution's aims, and indeed those of medicine in general. In theory the doctor's art had two primary components, preventive and curative, but the majority of physicians focused exclusively on the second. 'Preventive medicine', Beddoes mused to Tom Wedgwood, 'has never been cultivated, though it is much the most important of the two divisions'. Physicians 'had no motive to attend to it', because their livelihood consisted in treating the sick rather than the healthy. The healthy, by the same token, had 'never felt its importance – though I believe they may now be brought by proper measures to be sensible of it'.[7] By the end of 1797 Beddoes had relaunched his subscription drive for the Pneumatic Institute with a new pamphlet, but he had also begun to recast its mission in even more ambitious terms.

One of his most suggestive pieces of evidence that preventive medicine might have the power to reduce the incidence of consumption was that some classes and occupations seemed to be almost wholly resistant to it. Conspicuous among these, as Beddoes had noted during his years of practice, were butchers, a characteristically sleek, fat and rosy breed who rarely, if ever, seemed to display the sallow and drawn aspect of consumption. He began a survey of butchers in Bristol, and his anecdotal hunch proved accurate. One told him that he had 'never heard of a man dying of consumption who was a butcher', and added that 'after a sheep is dead, it is very wholesome to swallow the steam, the smell of meat keeps us from disorders'.[8] If this were true, the Pneumatic Institute might not after all be attempting to introduce a novel form of inhalation therapy, but rather to investigate and improve on one that had been practised in abattoirs and meat markets for centuries. A cure for consumption might be hiding in plain sight, literally under the physicians' noses, waiting only to be bottled and administered.

As he extended his researches from Bristol out into the neighbouring towns of Bath and Worcester, Beddoes was moving in parallel with another doctor who was also investigating folk beliefs that connected proximity to livestock with the prevention of disease. The previous May, Edward Jenner, a physician practising in the Gloucestershire countryside, had taken a tiny quantity of fresh pus from a sore on the arm of a dairymaid who was afflicted with smallpox and had scratched it into the arm of an eight-year-old boy named James Phipps; the boy developed a mild case of the disease that quickly receded. Six weeks later, in a bold experiment, Jenner had deliberately infected Phipps with a larger dose of diseased lymph matter, but smallpox had not followed. In 1797 Jenner repeated the experiment with three more subjects during a smallpox outbreak, with the same results, and submitted a report to the Royal Society, of which he had been a fellow since 1788. But Sir Joseph Banks was no better inclined to Jenner's experiments than he had been to Beddoes': the theory, he pronounced, was based on evidence too flimsy to warrant inclusion in the Society's *Transactions*.

Although medicine was absorbing Beddoes once more, there was a final contribution to the political debate that he still wished to make. His *Essay on Pitt* had been only the first volume of a projected pair: his chronicle of Pitt's rise to power had been left hanging with an advertised second volume, on Pitt's record as war minister, still to come. The first part of the *Essay* had been well received in the reform-minded press, with the *Monthly Review* hailing Beddoes as a 'bold and original thinker';[9] but he had been unhappy with Joseph Johnson's handling of it, which he felt had suffered from poor advertising and distribution: 'if I did not think him an honest man', Beddoes had muttered darkly to Giddy, 'I should suspect foul play'.[10] Now, the remainder of the polemic emerged from a new publisher, in a form designed to address the national mood that had taken hold since the passing of the Two Acts and the military resurgence of France under the Directory. *Alternatives Compared*, with its subtitle *What Shall the Rich Do to Be Safe?*, marked a shift in strategy towards his readership: if the propertied classes were reluctant to address the fate of the poor and starving on the grounds of compassion and charity, perhaps they might pay more attention to a book that promised to tell them how to save their own skins.

Beddoes' assessment of the state of play in 1797 was that 'the frantic paroxysm is past', the rush to the barricades defeated; the question now was: 'who can tell by what fatal torpor it will be succeeded?'[11] Starvation and riots, demonstrations, assemblies and petitions had made it abundantly clear that the government's policy was opposed by a vast swath of civil society, but that they were bent on pursuing it regardless. In this situation the alternatives for Britain were stark: either to continue attempting to force 'the unconditional surrender of the republicans', and plunge Britain into civil war; or to 'sit quiet, wishing that things might come round', or, perhaps, to 'bestir ourselves against the ministry with as much alertness as if we had to rescue all we hold dear from a building in flames'.[12] There was nothing to be hoped for by continuing down the current road. Britain had lost all her credit and prosperity; she was locked in an unwinnable war with no exit strategy; the 'precious inheritance which every Englishman derived from the exalted reputation of his country, is irretrievably gone'. None of this could be undone, and for most it could no longer be avoided; 'the wish for an asylum has crossed the mind of many a father, anxious for his family', [13] but apart from a mass exodus to America the only hopeful course was for Pitt to resign. His departure would drain the venom that was poisoning the nation, and indeed the man himself 'would be much more the object of compassion, and much less of hatred, in retirement than in power'.[14]

Beddoes' voice was not an isolated one: it reflected a public discontent that still seethed against a government that was painting itself into an ever tighter corner, and many were reacting to the Two Acts by supporting ever more drastic forms of resistance. On the brilliantly clear morning of 16 April 1797 the British fleet at Spithead had mutinied, refusing to obey their officers' orders until the repeatedly broken promises to improve pay and conditions were met, and even threatening to sail across the Channel and join the enemy. To Pitt's dismay, a large section of the public had supported the mutineers, and the opposition to the war had begun to rally around threats of mass military disobedience against the government. The mutiny had spread to the Thames, and had been put down only by the desperate efforts of double agents, hangings and yet another extension of the sedition laws, the Seduction from Duty Act. Unless a divided nation could somehow be reconciled, the alternatives were grim: submission either to a tyrannical government or to a tyrannical enemy.

* * *

But the impasse facing the Pneumatic Institute, at least, was about to be broken. Over the second half of 1797 the efforts Beddoes had expended on behalf of others would combine to reward his perseverance in unexpected ways.

The chain of events began with Tom Wedgwood's return from Germany. Staying with his brother John in Cote House, he made the acquaintance of Tom Poole, who suggested that he come to Nether Stowey to visit the young poet whom he had installed in a small cottage around the corner from him in the village. The rapport between Beddoes' two protégés was immediate; Tom Wedgwood's five-day visit with Coleridge passed 'like lightning',[15] and with consequences that would be far-reaching for all concerned. William and Dorothy Wordsworth were in residence at Alfoxden House, about three miles away, and the poets were in the legendary white heat of their communion: the verses piling up on Wordsworth's desk would soon emerge from Joseph

Cottle's press as the *Lyrical Ballads*, and they were shortly to take the walk over the hills to the cliffs of Porlock and Lynton during which *The Rime of the Ancient Mariner* would begin to take form. Wordsworth, who had been expecting Tom to be a wealthy bore, found instead a rare incarnation of the sensibilities that his poetry was seeking to awaken. He invited the party down to Alfoxden, and later wrote that Tom's 'calm and dignified manner, united with his tall person and beautiful face, produced in me an expression of sublimity beyond what I ever experienced from the appearance of any other human being'.[16]

For Tom, the idyll was the beginning of a lifelong friendship with Coleridge in particular, but also the trigger for a concerted dialogue with his brother Jos about how to dispense the benefits of the estate they had inherited. In Tom's view, the goal could only be that to which he had self-consciously dedicated himself years before, to improve the lot of mankind; like Beddoes, he believed that this would be most thoroughly accomplished by a reform of education, from which all other progress would flow. Tom and Jos had been toying with the idea of setting up an educational establishment, a 'Nursery of Genius', and their first interest in Coleridge and Wordsworth had been not as poets but as possible teachers. Tom's visit to the Quantocks, though, had immediately shown him that Coleridge was not a schoolmaster, but rather an incessant fountain of learning and philosophy whose ideal function would simply be to gush into the world at large, funded just to be himself.

Even though Coleridge was living rent-free thanks to Poole's generosity, he was desperately short of money: notes for ever more ambitious projects piled up and gathered dust while he laboured on 'reviews in the magazine and other shilling-scavenger employments'.[17] When Tom and Jos initially offered him a loan of £100, he refused: the all-too-likely prospect that the money 'would soon be consumed, and prospectless poverty recur' was enough to spark an anxiety more acute than any benefits he could imagine.[18] He craved a steady income, for which the most promising offer he had received was to serve as minister at the Unitarian Chapel in Shrewsbury. But Tom and Jos were not so easily put off. Coleridge, they believed, represented the genius of the coming generation, and the full flowering of his promise would be crucial to its hopes. Preaching, or indeed poetry, seemed to them inadequate vehicles for the scope of his vision. They continued to deliberate; and Beddoes, as well as Coleridge, would be a beneficiary of their eventual decisions.

*   *   *

In Bristol, meanwhile, Beddoes was taking his first steps to seed the idea of preventive medicine in the public mind by planning an open course of lectures on anatomy, through which the general public could learn the essentials that would empower them to take charge of their own health. 'Preventive medicine', he would tell his audience, 'has yet no existence': there was no money in it for the medical profession, and the public were at the mercy of the empirics and folk remedies that constituted their impoverished medical understanding. Its development would require 'the joint efforts of the intelligent in the profession and out of it', the skills and learning of the physicians, and the appetite and energy of the public.[19] By the end of the summer he had developed a plan for the course, which he sent to Tom Wedgwood for approval. 'I know very well that some at least of my chemical lectures at Oxford were dull', he confessed, 'even though I had greater classes than any lecturer before or since'; he hoped the new plan would make this scheme 'interesting and talked of in other places'.[20] By October he had issued a public prospectus, promising lectures that would 'exhibit the structure of the human body in a manner neither superficial nor tedious', and 'intersperse such reflections as may be useful in physical education, and the whole conduct of life'.[21] A committee, which included Tom Wedgwood and James Watt junior, advanced £50 for his expenses.

The introductory lecture for the course was set for November, with Beddoes commissioned to write it and Francis Bowles, who had been lecturing and working as a surgeon in Bristol for several years, designated to read it; but Beddoes, in Coleridgean manner, was still scribbling illegibly, a watch by his side, when Bowles was scheduled to begin. Bowles stumbled through the text with apologies to his confused audience, and Beddoes made amends by sending the text to Joseph Cottle to be printed. Though dashed off at breakneck speed, it was lively and cogent, and set out the prospectus for a programme that would occupy him for years to come. It began not as his works for physicians typically did, with a litany of medicine's ills, but with a catalogue of its recent successes designed to enthuse the public with its future possibilities. Scurvy was being checked (which, he now accepted, had been achieved by a diet of 'acid vegetables'); the danger of smallpox was being reduced by inoculation (a practice Beddoes still associated not with Jenner but with Lady Mary Wortley Montagu, who had discovered a similar practice in Turkey and popularised it to some extent in Britain in the 1720s).[22] But huge benefits were also accruing daily from a better understanding of the causes of disease. Better ventilation, less restrictive female dress, increased sobriety (but 'are we yet sufficiently temperate?')[23] were all improving constitutions, reducing contagion and aiding recovery. What was now needed was for the public to act systematically, and to become their own 'inspectors of health'.[24] In the process they would not only preserve their own lives, but rescue the medical profession from its 'lucrative indolence'.[25]

The full course of lectures, three a week for twenty-four weeks, began on New Year's Day 1798. It was well attended and widely discussed: Beddoes' conviction that there was an untapped demand for medical education among the public was vindicated. The lectures made a welcome change, too, from solitary writing and arduous projecting, and reminded him how much he had missed the crowded lecture halls of his Oxford years. When the lectures concluded, he ran a second series specially adapted for a female audience, the first time that women had ever been offered public education in Bristol. Forty 'ladies of great respectability' signed up for a series of ten lectures on physiology, bones, the sense organs, and the heart and circulation, from which all indelicate material had been carefully excised.[26]

Beddoes' medical practice was busier than ever, and the canvas of his life crowded with tragic and desperate cases of consumption. He was much preoccupied with a young lady whom Davies Giddy had referred to him, Lydia Baines, who was stricken with hectic fevers and wasting away rapidly. Beddoes tried the airs on her, but 'whether it was an accidental coincidence or not, the cough for those 2 or 3 days was much aggravated'; he switched to ether, a volatile inhalant favoured by Darwin for constricted chests, which provided only temporary and symptomatic relief.[27] As a last resort he decided to experiment with an idea that had struck him while he was researching the resistance of butchers to consumption: 'to imitate the exhalation of a cow-house'. The breath of cows was low in oxygen, since the cows had already sequestered it into their bodies, and was thus a natural approximation of the factitious treatments for consumption he was developing in the laboratory; but he was also 'quite persuaded from experience of the power of those fumes to give a healing stimulus to ulcerated lungs in some cases'. It was still unclear to Beddoes 'what it is in that medley of elastic matter which does service', but he suspected that somewhere in the heady steam of meat and milk lay a remedy as yet undiscovered.[28]

But the most significant of his patients, as it turned out, was William Lambton, a wealthy coalmine owner from Durham who had been directed to him by William Reynolds. Lambton had suffered from consumption for many years, and had now reached the all-too-familiar point of last resort. During the autumn Beddoes had urged him to travel south for the winter, and by the end of the year he was receiving entreaties from Lambton 'in a manner not to be resisted, to set off to Italy'. Beddoes had virtually decided to join him in Naples when he received the news that he had died.[29]

Beddoes, who had become extremely fond of Lambton, was deeply distressed, not least because he had left five young children behind to cope with the tragedy. He collapsed into a condition that he would come to refer to as his 'Hamlet's complaint', in which his usual good cheer and bustling activity gave way to a consuming melancholy.[30] It was a mood that would return to dog him in later life; its onset at the end of 1797, though triggered by loss, guilt and helplessness at failing to save a well-loved patient, may also have reflected his exhaustion at projecting the dream of an Institute that seemed to be for ever on the horizon. Part of him, perhaps, had come to wish that it might remain there: he had often, particularly in correspondence with Giddy, anticipated its failure, or attempted to persuade himself that it made no difference whether it went ahead or not ('if the Institution should not be established the proposal and advertising will have been beneficial').[31] But if despair was threatening to take up residence in his soul, it was soon to be driven away. The two remaining obstacles to setting up the Pneumatic Institute, funds and a suitable laboratory superintendent, were both shortly to be overcome.

*   *   *

Among the procession of consumptives travelling for their health in the winter of 1797 was James Watt's younger son, Gregory, to whom Erasmus Darwin had recommended the mild sea air of the south-west of England. Beddoes had referred him to Davies Giddy to find accommodation, and Giddy had recommended that he should lodge with a widow in Penzance named Grace Davy who offered bed and board.

Mrs Davy was the mother of five, of whom the eldest male, Humphry, had recently come to Giddy's attention while swinging from the garden gate of John Bingham Borlase, the surgeon-apothecary to whom the seventeen-year-old had been apprenticed for two years. Giddy's companion had pointed Humphry out as the son of the woodcarver Robert Davy, who had died three years previously, and mentioned that he was fond of making chemical experiments. 'Chemical experiments!' Giddy had exclaimed. 'Then I must have some conversation with him.'[32] He did so, and quickly discovered that young Davy was something of a prodigy. His interest in chemistry had begun in childhood, when he had made fireworks with his own recipe of 'thunder-powder' and ingots of tin by melting down fragments with a candle in a hollowed-out turnip. During the last two years, however, he had begun to pursue the science systematically and voraciously, through the limited means at his disposal. He had learned French from an exiled priest from the Vendée who had taken up residence in Penzance; though he had shown no interest in attempting to speak it, he had quickly mastered reading and had plunged directly into Lavoisier's *Traité élémentaire de chimie*, which he read alongside a recent English textbook, William Nicholson's *Chemical Dictionary*. Comparing and contrasting these two sources, he had developed both a thorough understanding of the subject and a thrilling sense of the conflicts that were waiting to be resolved.

Giddy invited Davy over to Tredrea to use his library, and he became a regular visitor. His favourite aunt lived close by at Marazion, the small harbour that faced Penzance across the sweep of Mounts Bay, to which Davy made the three-mile walk along the coach road beside the sandy shore as often as he was able. He quickly demonstrated an appetite and aptitude for learning unlike anything Giddy had previously witnessed. 'I know of few things more out of the common way', he told Jos Wedgwood, 'than a young man in a remote country town, without advantage and almost without books or apparatus, carried forward by the force of his own genius'.[33] Without access to laboratory equipment, Davy had somehow contrived to become a sophisticated experimentalist, pressing into use whatever instruments he could find. He had investigated the composition of the airs inside seaweed bladders, and had established that seaweed releases oxygen in the presence of light, just as Priestley and Darwin had shown that plants did on land. On another occasion he had found a rusted enema syringe among the contents of a shipwreck, and had customised it to make an air pump; with this, he had performed an ingenious experiment with a clockwork device resting on a bed of ice, pumping out a vacuum to demonstrate elegantly that the ice, even when denied air or heat, would nevertheless melt from the vibration generated by mechanical motion. When Giddy took him to visit a friend who had at his home 'a quantity of chemical apparatus hitherto only known to him through the medium of engraving', Davy's delight was indescribable: he put the air pump through its paces 'with the simplicity and joy of a child engaged in the examination of a new and favourite toy'.[34]

Although he had always shown a remarkable quickness in things that interested him, and from an early age had had the ability to flip through a book as if looking for pictures and then parrot its contents in remarkable detail, Davy had made it through his early life without being singled out as particularly bright. His enthusiasms were not for schoolwork but for play: fishing, sketching, and composing fantastical stories and verses. In 1793 he had been sent to the county grammar school in Truro, where he had absorbed the rote learning of the curriculum adequately but with no great appetite or distinction. He had shown the most flair at verse translations, in which his teacher, the Rev. Dr Cardew, had encouraged him.

1 Bastille Day 1791: Joseph Priestley burned in effigy by drunken rioters outside the Dissenters' Meeting House in Birmingham, together with revolutionary texts including Thomas Paine's *Rights of Man*. The rioters proceeded to destroy Priestley's home and his priceless chemical laboratory.

2 Thomas Beddoes M.D.

3 Davies Giddy at his home in Tredrea, near Penzance, Cornwall.

4 Tredrea today.

5 The world's first iron bridge, constructed in 1773 over the river Severn in Shropshire from 375 tons of iron, cast at the nearby Coalbrookdale works.

6 Coalbrookdale *c*.1800 by Phillip de Loutherbourg, who described it as 'worthy of a visit from the admirer of romantic scenery, no less than from the political economist'.

7 William Reynolds, ironmaster of Ketley and Beddoes' patron in Shropshire, in his habitual Quaker dress (1796).

8 Erasmus Darwin, Beddoes' mentor as physician, philosopher and poet, painted by Joseph Wright of Derby in 1770.

9 Hotwells in 1791, in a watercolour by the sixteen-year-old J.M.W. Turner. The hot springs, boarding houses and promenade cluster under the spectacular scenery of the Avon Gorge.

10 Richard Edgeworth.

11 Tom Wedgwood.

12 Anna Beddoes.

13 Beddoes and Anna's home in Clifton at 3 Rodney Place, in a small crescent off the village high street.

14 and 15 Samuel Taylor Coleridge and Robert Southey during their turbulent residence in Bristol, 1795–6.

16 The new Bristol Bridge, site of the violent riots of 1793.

17 James Watt, whose daughter Jessie and son Gregory were both treated for consumption by Beddoes.

18 and 19 James Watt junior, son of the great inventor and notorious 'English Jacobin', who attempted to enlist support for the Pneumatic Institution from the President of the Royal Society, Sir Joseph Banks (below).

20 Humphry Davy as a young man.

21 As Sir Humphry, President of the Royal Society.

22 St. Michael's Mount in Mount's Bay, Penzance, the magical horizon of Davy's youth. This engraving was published by the Cornish antiquary William Borlase in his pioneering survey of the county's ancient sites (1769).

23 The site of the Pneumatic Institution in Dowry Square, Hotwells. The Institution was at Nos. 6 and 7, the corner of the terrace, with workshops, laboratories and storage in the yard and outhouses at the back.

24 James Gillray's satirical print of Davy's nitrous oxide demonstration at London's Royal Institution on 20 June 1801. Davy is depicted as a rustic youth holding a set of bellows and sniggering as Count Rumford offers the gas to the distinguished spectators in evening dress.

25 Gillray did not single out the nitrous oxide researches for ridicule. Here he lampoons another scientific miracle story of the day, Edward Jenner's new vaccination therapy (1802).

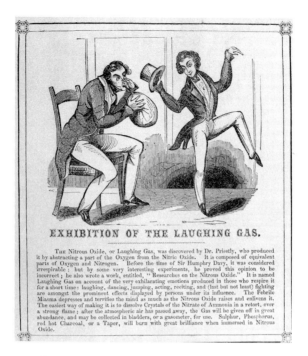

EXHIBITION OF THE LAUGHING GAS.

The Nitrous Oxide, or Laughing Gas, was discovered by Dr. Priestly, who produced it by abstracting a part of the Oxygen from the Nitric Oxide. It is composed of equivalent parts of Oxygen and Nitrogen. Before the time of Sir Humphry Davy, it was considered irrespirable: but by some very interesting experiments, he proved this opinion to be incorrect; he also wrote a work, entitled, "Researches on the Nitrous Oxide." It is named Laughing Gas on account of the very exhilarating emotions produced in those who respire it for a short time: laughing, dancing, jumping, acting, reciting, and (last but not least) fighting are amongst the prominent effects displayed by persons under its influence. The Febrile Miasma depresses and terrifies the mind as much as the Nitrous Oxide raises and enlivens it. The easiest way of making it is to dissolve Crystals of the Nitrate of Ammonia in a retort, over a strong flame; after the atmospheric air has passed away, the Gas will be given off in great abundance, and may be collected in bladders, or a gasometer, for use. Sulphur, Phosphorus, red hot Charcoal, or a Taper, will burn with great brilliance when immersed in Nitrous Oxide.

26 During the nineteenth century, nitrous oxide became a popular entertainment and earned its nickname of 'laughing gas'. Humphry Davy's bold experiments and the high-flown pronouncements of the poets combined to set the tone for public 'laughing gas' frolics.

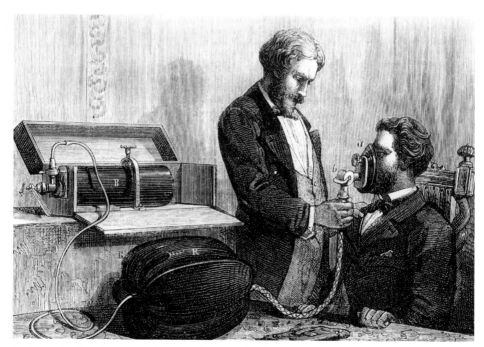

27 Observing these frolics gave dentists and doctors the idea of using nitrous oxide to remove pain in surgery. By the end of the nineteenth century it was in common medical use and the term 'anaesthesia' had been coined to describe its effects.

With the advent of his teenage years, poetry had become Davy's passion; or, perhaps, the medium through which he distilled a deeper passion for the wild landscape of his native countryside. Growing up in Penzance, Davy had lived his whole life in the roar of the wind and sea, and before the constant spectacle of nature at its most grandiose. Surf, spray and sea mist refracted sun and moon in an ever changing show; each day brought a dawn and sunset of panoramic intensity, each different from the last. The phases of the moon, haloed by frost, rain or salt spray, cycled the tides through a vast reach, revealing sunken expanses of reef and strand; the breakers and deep swells spoke of forces that were building and discharging far beyond the horizon, and in the ozone smell of the wind and sea spray he could sense these invisible forces coursing through his own system. Davy loved nothing more than to lose himself in this elemental drama where the protagonists of nature – rocks and trees, wind and water, sun and moon – played out their infinite repertoire. On his favourite bays and head-lands, where these forces were at their most intense, he could feel himself part of them; and it was from these harsh and magnificent scenes that he drew his poetry: where, as he put it,

Enthusiasm, Nature's child
Here sung to me her wood-songs wild
All warm with native fire.[35]

Meeting Giddy awoke for the first time in Davy the sense that he might be destined for greatness. His father had been unambitious, and not good with money; from an early age his mother had looked to the man who had brought her up, a local surgeon-apothecary named John Tonkin, to advance her son's prospects. It was Tonkin who had paid Davy's school fees and, when his father died, had apprenticed him to his colleague John Borlase. Tonkin was an elderly, old-fashioned man who still dressed in powdered wig, buckles and tricorn hat; he took his responsibilities as guardian seriously and knew that he had set Davy, for a Penzance boy with no means, on the most reliable career path open to him. The city had a long-standing reputation as a rough outpost at the ends of the earth, whose inhabitants colluded routinely in smuggling and wrecking and dissipated the proceeds in drinking, gambling and cockfighting; but in recent years, thanks largely to the inroads made by Methodism, it had begun to improve itself, and now boasted a book club, a theatre and even a civic Assembly Room. Tonkin was determined that Davy should become an exem-plar of the city's self-improvement, and that his apprenticeship should lead to a medical diploma, his passport to the ranks of the local professional class as a physician.

Borlase ran a very busy dispensary, prescribing treatments for the rheumatism that was endemic in the damp climate, the occupational hazards of miner's lung and the emergencies of pit accidents; but Davy, though he liked the sound of a medical career, showed little interest in the chores of apothecary's assistant. His scientific curiosity was at this point focused on geology, the stuff of the nature that he loved. He was fascinated, as Beddoes and Giddy had been over the summer of 1791, by the variety of the granites and the lodes and dykes of glittering metal and crystal that shot through them, about which he gleaned all he could from the miners who, together with fishermen and smugglers, made up most of his community. His regular walks along the shore to Marazion and Tredrea were his geological expeditions, when he could be found swinging a hammer and filling his pockets with specimens. In the words of his first biographer, 'he thought more of the bowels of the earth than the stomachs of his patients', and 'when he should have been bleeding the sick, he was opening veins in the granite'.[36]

Giddy had been looking forward to introducing the delicate but brilliant Gregory Watt to 'the extraordinary person from Penzance',[37] but the two young men proved slow to warm to one another. Watt, like his father, hid a generous heart behind a diffident and cool exterior; Davy, by contrast, had a wild, almost feral quality that was alien to the sober and genteel Birmingham milieu in which Watt had grown up. It was typical that Giddy had first caught sight of him swinging on a gate: Davy spent as much time as possible scrambling up cliffs, wading through brooks and tramping across moors, always covered in cuts and bruises that he rarely noticed, insisting that any pain could be made subservient to firm mental discipline. The worship of wild nature in his poetry was no affectation: clumsy as its expression may sometimes have been, the sentiments were profoundly felt. When roused from his solitary obsessions, he was open and ebullient company and, excited by a distinguished guest, overpowered the delicate Watt with bumptious enthusiasm and wildly declaimed poetry.

But relations between the two thawed as soon as Davy turned the subject to chemistry, and began to reel out his progress in the subject since Giddy had taken him under his wing. Watt was astonished to hear the uncouth rustic lad not only dismissing the phlogiston theory but insisting that there was equally little evidence for the existence of caloric: Lavoisier held air to be the source of caloric, whereas Davy's own experiments had shown that heat could be generated in a vacuum, demonstrating that it was not an element but an effect of motion. Before long the invalid Watt was scampering after Davy up cliffs and through rock pools, mapping his superior theory onto Davy's store of local sites, both men returning heavy with mineral samples. Watt was, along with Beddoes, a pioneer in adducing chemical processes to support Hutton's geological theories, and he set Davy's mind racing on new tracks. He also discovered Davy to be naïve in civilised pleasures, and later recalled the memorable occasion 'when I had the honour of being your mystagogue in your initiation into the orgies of the mirth-inspiring Bacchus': Davy, staggering drunk for the first time, had attempted to howl and stumble through his paean to St Michael's Mount above the roar of the surf.[38]

At the end of January, Watt was joined in Penzance by Tom and Jos Wedgwood, also in search of a mild winter destination for Tom's health. The meeting between Tom and Davy was a fortunate one. Davy's raw passion for chemistry was thus far unrefined by laboratory discipline; Tom, no longer able to undertake the painstaking experimental work of his teenage years, was glad to pass on his finely honed skills, and Davy would record in later years that Tom's advice had been a 'secret treasure' to him throughout his life, often teaching him 'to think rightly when perhaps otherwise I might have thought wrongly'.[39] Fresh from his acquaintance with Coleridge, Tom had found another genius in the making, and Davy listened eagerly to tales of Coleridge and Wordsworth to add to those that Giddy had told him of his Oxford mentor Dr Beddoes and his revolutionary pneumatic project in Bristol. A career as a Penzance apothecary, until a few months previously the summit of his ambitions, began rapidly to be replaced with grander horizons.

As Davy's grasp of his possibilities grew, so did a vague but potent sense of his destiny. As he recalled later, 'I gradually became conscious of my powers, by comparing them with others'';[40] it was only when Giddy opened the library and laboratory doors to Davy that he realised how fast he was able to master the sciences, and how greatly his speed exceeded the norm. As a youngster making explosions in John Tonkin's garret, he had jokingly been nicknamed 'the philosopher' and even, prophetically, 'Sir Humphry'; now he realised these epithets might be within his grasp, and he even headed a page of his notebooks 'Newton and Davy'. His discovery of his own unfurling talents raised him often to the pitch of ecstasy, producing the sublime sensations that he had thus far only experienced in nature and fleetingly captured in poetry. One of his early poems (and the first to make its way into print), 'The Sons of Genius', had already mapped the destiny that was now beginning to reveal itself to him. In it, the goddess Reason rises above superstition, creating a 'calm empire of the peaceful mind' where,

Inspired by her, the Sons of Genius rise
Above all earthly thoughts, all vulgar care
Wealth, power and grandeur they alike despise
Enraptured by the good, the great, the fair.[41]

This invisible college of like-minded scientists, unlocking the secrets of nature and transforming the world in the process, had been conjured as a daydream; now it was a serious aspiration, and one that was shortly to become a reality.

*　*　*

Just before Tom and Jos left for Penzance on 26 January 1798, they had finally brought their delicate negotiations with Coleridge to a mutually acceptable conclusion. They had offered him an annuity of £150 a year for life, removing his guilt about lapsing into debt once it was spent and allowing him to express himself free from the necessity of scrabbling for journalism or living up to the expectations of a church. Although William Hazlitt remembered that Coleridge 'seemed to make up his mind with this proposal in the act of tying on one of his shoes',[42] it had taken a great deal of agonising – and some firm urging from Tom Poole, who had already done his share of supporting a financially embarrassed Coleridge – before he accepted. His first use of the Wedgwood annuity would be to travel to Germany to study with professors such as Johann Friedrich Blumenbach whose work he had encountered in the library at Rodney Place.

At the same time Tom Wedgwood cut the Gordian knot of the Pneumatic Institute's funding by offering a donation of £1,000, virtually doubling the level of subscriptions and bringing them to a total where the project could finally be realised as planned. Ever since subscriptions had opened, it had been expected that Tom would contribute; the only question had been how much, and when. Initially he had held back, wary of the danger of 'deterring others by too large a subscription',[43] and waiting to see how much could be raised from a broad base of small contributors. Now, though, it was clear that this first wave of subscriptions had raised a sum that was respectable but not quite sufficient; with the death of their father, the commitment that the Institution represented had shrunk to well within Tom and Jos' means. Beddoes had shown all the energy, ingenuity and persistence that could have been expected of him; now, the airs would have their chance.

Finally the Medical Pneumatic Institution had a schedule: it 'will be set on foot in winter', Beddoes told Giddy, 'if times permit'.[44] He could not entirely shake off his ingrained caution that the end of the year was too far ahead to plan for. He was right that turbulent times lay ahead: within a month, the government would close down the corresponding societies, round up dozens of suspected opposition figures and imprison them without trial; within two, the Great Rebellion would have broken out in Ireland, setting off a chain of insurgency and repression that would leave Britain more vulnerable to attack than ever. But Beddoes had waited long enough for the times to change before he could begin his work. 'I am writing down all the variety of volatiles I can think of', he told Giddy, 'to try in the Pneumatic Institution'.[45] He was in the middle of a series of chemical lectures in Bristol which he had prepared to follow up the success of the anatomy course, reprising some favourite experimental demonstrations of his Oxford days, and there was even talk of opening up a public scientific institution in the city on the wave of enthusiasm he had generated. But from this point these side projects would begin to drop away. The Institution's moment had finally arrived.

Now that he had funds, he had two urgent requirements: new premises, and a laboratory superintendent. For the latter, Giddy already had a recommendation. In April, Beddoes received a package from Penzance containing three papers from Humphry Davy, setting out his experiments and conclusions on the nature of heat and light. He claimed to have succeeded in passing beyond the compromise solution that, both for him and for Beddoes, Lavoisier's theory of caloric represented. Caloric was a weightless spirit within fire that dissolved matter, and it had led Lavoisier to describe oxygen in ambiguous terms as a 'principle' as well as an element; Davy proposed that we should rather speak of 'phosoxygen', a compound term that captured the twinned roles of light and oxygen in transforming matter and animating life. While heat could be reduced to motion, light, he theorised, was a driving force in chemical reactions. It was composed of infinitesimally tiny particles, which are attracted by the particles of ordinary matter in ways that obey Newton's laws of optics and refraction; but they also, when they enter our bodies, feed our blood and stimulate our nervous systems in ways that Brown and Hartley had sought to describe, acting on our organs of vision and fuelling our systems of perception. Thus light is the prime mover of both the material world and the mental, connecting us all in a vast web of life that reveals 'the sun and the fixed stars, the suns of other worlds, as immense reservoirs of light destined by the great Organiser to diffuse over the universe organisation and animation'.[46]

The essays, running to over two hundred pages and ranging from theories of heat and light to phosphorescence and respiration, wove experiment and speculation into a cosmic vision that, as the product of an untrained eighteen-year-old, impressed Beddoes hugely. He too was looking for a new way of understanding heat and light, and reluctant to follow Lavoisier. 'I have some time ago given up caloric', he wrote the same month that he received Davy's papers, 'and have thought that the phenomena are well reducible to vibration'.[47] Davy's theories were particularly congenial to him as they fed back his own thoughts, which Davy had gleaned from Giddy's library, to meld the principles of chemistry with Brunonian medicine, positing light as the source of the excitability that drives both chemical reactions and living metabolisms.

If Davy had embroidered his experimental findings a little too enthusiastically, his speculations nevertheless seemed to Beddoes to be leading in promising directions, and ones that chimed impressively with those of the most brilliant theorists of the day. Beddoes was simultaneously reading the new paper on heat by Count Rumford, reported to the Royal Society in January, which also argued that 'percussion or friction can produce heat without change of capacity or oxydation'.[48] Rumford had made his observations while working in a cannon foundry in Munich by harnessing horses to a massive engine that, through a gearing system, turned a lathe that heated the brass as it bored the cannon barrel; Davy had achieved the same result two years previously using a cup, some clock workings and a syringe recovered from a Cornish wreck. Beddoes forwarded Rumford's paper to Davy, and asked permission to print his essays in a volume he was planning.

It was obvious that Davy was an outstanding candidate for the post of laboratory superintendent. Beddoes sounded the possibility out with Giddy, who had already made the same recommendation, but had warned that Davy was currently bound by the terms of his apprenticeship to Borlase, and to disengage him would require delicate handling. Beddoes argued that to accept the post was clearly in Davy's own best interests: 'I think I can open a more fruitful field of investigation than anybody else. Is it not also his most direct road to fortune?'[49] But the fortune would not come immediately: 'he must be maintained, but the fund will not furnish a salary from which a man may lay up anything'. Giddy passed on Beddoes' offer, and communicated by return that Davy requested no more, but no less, than a 'genteel maintenance'. 'I can attach no idea to the epithet "genteel" ', replied Beddoes, 'but perhaps all difficulties would vanish in conversation'. Davy was not, he suggested, abandoning his career path: 'this appointment will bear to be considered as part of Mr. Davy's medical education'.[50] He was simply trading one apprenticeship for another with prospects infinitely brighter.

Giddy communicated the offer to Borlase, and the apothecary agreed to release Davy from his apprenticeship with a legal certificate stating that he was doing so because 'being a youth of great promise, I would not halt his present pursuits, which are likely to promote his fortune and his fame'.[51] But John Tonkin, Davy's benefactor since childhood, was not so easily won round. The Pneumatic Institution seemed to him a very uncertain venture, one that was as likely to bring disaster as success, and disgrace as fame. He may, as others in Penzance certainly did, have recalled Beddoes' unfortunate influence on Giddy, and their pamphlet attacking the idea of a Cornish county infirmary in 1791. Davy insisted that he still intended to study medicine at Edinburgh; but if this was the case, the position on offer was surely a pointless and risky diversion. A physician's diploma could be counted on to supply a decent living and position in society, but a project supported by private subscription was a gamble, a house of cards that might fold at any moment. After Davy's departure Tonkin redrafted his will, striking out the clause in which he had left his house to his young ward.

* * *

As soon as these negotiations were concluded, Beddoes was summoned unexpectedly to Durham by William Lambton's widow. She had told him that she wanted his help in laying down medical and educational plans for her five children, but on his arrival she revealed that she had a further request: she wished to entrust the two eldest boys to his care until they were ready to be sent to school at Eton in four years' time. They would come with a generous allowance for their maintenance and also, in memory of their father's untimely death and in hope of a future cure for his condition, a subscription of £1,500 to the Pneumatic Institute.

Beddoes was initially shocked by the request, but quickly recognised that it was as sensible a solution as any: 'I am very far from thinking that we can come up to ideal perfection in educating these boys; but who can?'[52] He was, at least, not short of educational theories with which to experiment. Richard and Maria Edgeworth's long-gestating collaborative work on the subject, *Practical Education*, had appeared that year, to immediate acclaim: it would remain a popular primer for decades to come, not only in Britain but across the Continent in Dutch, Italian and German editions. It incorporated much of Beddoes' thinking, especially on rational toys, into a scheme that encouraged the replacement of rote learning with practical exercises and the freedom to experiment, all governed by a theory of association aimed at improving attention, memory and judgement. Controversially, it omitted religious instruction entirely from its programme: in later editions Richard Edgeworth justified this exclusion on the grounds that most parents already had their own strong views

on religion, and plenty of tracts and primers to guide them: 'Can anything material be added to what has already been published on this subject?'[53] Maria, for
her part, claimed 'for my father the merit of having been the first to recommend,
both by example and precept, what Bacon would call the experimental method
in education'.[54]

The two Lambton boys, aged seven and five, moved into Rodney Place, where
Beddoes and Anna treated them as members of the family: Anna would later
confess that she came to wish they were her own. Their education took place at
home, where they were allowed to play as they wished, with the proviso that they
avoided idleness and maintained 'a constant state of happy activity'.[55] In the
morning Beddoes worked, leaving them to their own devices; in the afternoon he
read simple and classic literature with them, practised conversation and debate, or
turned them over to one of the constant stream of learned visitors for instruction
in chemistry or anatomy. They learned mechanics from the rational toys that he
provided, and as they grew older he took them with him on his travels. Unlike
many well-intentioned Enlightenment experiments with education, both the experience and the eventual outcome were happy for all concerned. Anna had a full and
bustling house during Beddoes' absence, and Beddoes 'never appeared to enjoy
himself more completely, than during this period'.[56] The boys would arrive at Eton
conspicuously well educated, mature and independent, and go on to successful
adult lives: the elder, John George, as the first earl of Durham, would earn the nickname 'Radical Jack', and play a significant role in drafting the Reform Bill of 1832.

No sooner were the 'Lambkins' installed than a new lodger was on the way.
On 2 October 1798 Humphry Davy boarded a coach with Davies Giddy in
Penzance and two days later crossed the border into Devon, the first time that
Davy had ever left his home county. They stopped at the market town of
Okehampton, its forbidding granite streets huddled beneath the high bluffs of
Dartmoor, to find the population in high spirits: the mail coach had just arrived
from London, crowned with laurel and streaming with ribbons, bringing news
of Admiral Nelson's victory over Napoleon on the Nile. All the rest of the route,
through Exeter and Taunton to Bristol, Davy found the people celebrating and
the towns illuminated through the night. When the Great Rebellion had broken
out in Ireland in May many, like Beddoes, had expected an invasion before the
end of the year; but Napoleon had sailed his fleet to the Mediterranean only
days before the news of the Irish insurgency had reached Paris. Now, thanks to
Nelson's brilliance, the British had pursued and destroyed the French fleet, and
perhaps turned the tide of the entire war in Britain's favour. It was an auspicious
moment for Davy's arrival, and for his own pursuit of destiny.

When Davy reached Bristol and met his new host, guardian and employer, his first impression was astonishment at his appearance. 'Between you and me', he confided in his first letter to his mother, 'Dr. Beddoes . . . is one of the most original men I ever saw – uncommonly short and fat, with little elegance of manners, and nothing characteristic externally of genius or science'.[57] Beddoes was stern, businesslike and perhaps nervous; he struck Davy as brusque, 'extremely silent, and in a few words, a very bad companion', though by no means unfriendly: he 'paid me the highest compliments on my discoveries, and has, in fact, become a convert to my theory'.

His first impressions of Rodney Place were considerably brighter: 'capacious and handsome', his own rooms 'very large, nice and convenient' – and, most welcome of all, an 'excellent laboratory'. If Beddoes had seemed forbidding, Anna made precisely the opposite impression: 'extremely cheerful, gay and witty; she is one of the most pleasing women I have ever met with. With a cultivated understanding and an excellent heart, she combines an uncommon simplicity of manners.' Only five years older than Davy, she was a maternal older sister, softening the texture of a world in which he had envisaged a punishing work schedule in the company of professional men much more senior than himself. He was delighted to discover that there was to be bright mixed company, and poetry to leaven the science.

After the bustle, filth and poverty of Bristol, Clifton was a surprise and a delight. 'Houses, rocks, town, woods and country in one small spot', he gushed to his mother, all perched above 'the sweet flowing Avon'.[58] In Cornwall, the only world he had known, human habitation had to be gouged out of an unforgiving landscape and battened against the fierce weather; here, in the sheltered cleft of the Bristol Channel, construction was expansive and elegant. Fine town houses perched beside sheer cliffs, gardens abutted onto slopes verdant with exuberant vegetation, and the prospects of wild nature were cultivated rather than shunned. But Clifton would never inspire Davy to poetry as the magnificent desolation of Cornwall did; rather, it was the human landscape that fascinated him with the richness and beauty of its forms. He renewed his bosom friendship with Tom Wedgwood, who was once more resident with brother John at Cote House but often to be found at Rodney Place, teaching the children or tinkering in the laboratory. Together they journeyed out to Nether Stowey, where Davy made firm friends with Tom Poole and experimented with new chemical processes in his tanning yards. Through Wedgwood and Poole he heard much of the meteoric Coleridge who had streaked off to distant Germany, to return who knew when.

Davy also made a swift and deep connection with Robert Southey, who had returned to Bristol over the summer. After his rupture with Coleridge, Southey had travelled to Spain and Portugal with his uncle before attempting to knuckle down to a law degree in London; but poetry had remained his driving passion, and Bristol his heartland. He was now scraping a living by writing and reviewing (including a bitter rubbishing of *Lyrical Ballads* in the *Critical Review*), and renting a house with his wife, Edith, in the outlying village of Westbury. Davy found in Southey a brother in poetry, his first friend who was equally committed to the art and determined to make a living from it. Southey felt that Davy's talent was still rough-hewn, but he was an exciting discovery, and they discussed collaborating on a heroic poem: one with Indian or perhaps Zoroastrian themes, and plenty of scope for sweeping landscape descriptions. Southey was beginning to assemble contributions for a volume of poetry, and it would be in his *Annual Anthology* of 1799 that Davy's poems, 'Sons of Genius' and 'On Mount's Bay', would first see print.

Richard Edgeworth was quick to recognise a genius for invention similar to his own, and his ideas, advice and encouragement rapidly 'possessed much influence over Davy's mind'.[59] Davy also made the trip to Birmingham to visit Gregory Watt, who was astonished to encounter not the previous winter's wild Cornish lad but a confident and extrovert figure at the hub of a sophisticated philosophical circle. Yet within Davy, for all the polish of his new society, was something that remained untouched: a 'solitary enthusiasm', as he called it in his notebooks,[60] an independence that set him outside the gaiety and company even while he was apparently lost in it. Joseph Cottle claimed to identify in the intervals in his animated conversation a strange introversion 'amounting to absence, as though his mind had been pursuing some severe trains of thought, scarcely to be interrupted by external objects'.[61]

Among Davy's friendships in Clifton, his relationship with Anna burned with particular brightness. She delighted in showing him round the town, excursions that Davy would recall vividly in later life; they discussed poetry avidly, and Davy copied Anna's poems carefully into his own notebooks. He made some tentative jottings of his own beside them, in an open, lyrical style quite unlike his brooding nature poetry ('Anna thou art lovely ever/Lovely in tears/In tears of sorrow bright/Brighter in joy').[62] She was the first woman he had known with whom he could share his sensitivity to poetry, and through it his most private feelings: if she had become elder sister and mother to him, she had also become the object of a romantic crush. It was a species of attention of which Anna's own marriage offered all too little, and which she was prepared gently to indulge.

All these new friendships, however, took second place to the intense programme of work into which Davy and Beddoes plunged almost as soon as he had unpacked. They had to establish the best methods for synthesising all the factitious airs, measuring and raising their levels of purity and improving their delivery systems, and deciding which treatments were simple enough for everyday use and which would be more complex and demanding; but both were also eager to follow up the programme of work that flowed from Davy's papers on light and heat. 'When I left Penzance, I was quite an infant in speculation, I knew very little of light and heat', Davy wrote home to Giddy.[63] Beddoes, both skilful chemist and generous teacher, was drawing experiments and discoveries out of his student prodigy at an extraordinary rate. A fine vein of strontium sulphate had been found in rocks near the mouth of the Severn; Davy ran a series of experiments to determine its properties, and succeeded in producing a previously unknown salt by making it react with oxygenated muriatic acid (which, within a few years, he would reveal as an element and christen 'chlorine'). 'This salt possesses most astonishing properties', he told Giddy: when combined with sulphuric acid, for example, it gave off a spectacular luminescence.[64] It was a gruelling programme, but also a freewheeling one, where Beddoes allowed one experiment to lead to another, and a new train of thought to set a new direction. Ever suspicious of theories, including his own, Beddoes was aware that he had a practical prospectus to fulfil; but he also believed in allowing its course to be flexible to what Priestley called luck, but Davy would characterise as genius.

*   *   *

The bequest from the Lambton estate had put Beddoes in a position to overcome the Pneumatic Institution's final obstacle: the purchase of a property suitable for the complex of laboratory, workshop, outpatient surgery and residential hospital that the project had grown to require. He had identified a good possibility just before Davy's arrival, and negotiations for it were moving towards a satisfactory conclusion. It was a modern town house in Hotwells, standing at the top corner of Dowry Square, another recent development that, like Hope Square, had been built hard against the steep slope that led up to Clifton. It was separated from the Avon Gorge by the Hotwells Parade, but perched on enough of a rise to catch the breeze, and from its top storey the view to the wide expanse of estuarine river and rolling hills beyond. It was at the corner of a terrace, its tall and plain frontage concealing a back yard with a cluster of outhouses that would offer ample space for deliveries, laboratory workshops and storage for chemicals and gases. There would even be room for Davy to have his own private quarters on the top floor. 'By the ardent and incessant exertions of Dr. Beddoes', he announced in his notebook at the end of 1798, 'the Pneumatic Institution is at length on the point of establishment'.[65]

Alongside all the work required to launch the Institution, Beddoes was busy with publications that he wished to get to press before it opened. In August he had issued a prospectus for a regular series of publications entitled *Contributions to Physical and Medical Knowledge Principally from the West of England and Wales*, which was to form an evolving collection of reports, case studies and experiments from Bristol and its environs. It was conceived partly as a clearing-house for local medical researchers to share their discoveries, and partly as a showcase and platform for projectors to attract interest and funds. Its proceeds were to be donated to deserving medical institutions in the area, with those from the first volume benefiting the Pneumatic Institute.

As it turned out, the first volume was also to be the last, but not for lack of interesting material. It included among its miscellany a correspondence about the theories of Edward Jenner, whose first major publication, *Inquiries Concerning the History of the Cow-Pox*, had been published the previous winter. Despite the Royal Society's lack of interest, trials of Jenner's procedure – christened 'vaccination', to distinguish it from older practices of inoculation which involved dried rather than fresh lymph and were more uncertain – had spread rapidly through the region and were now being conducted as far away as Manchester. Many were finding the results impressive, but Beddoes remained unconvinced: he had observed folk practices of inoculation for many years and had seen plenty of cases to contradict Jenner's theory that cowpox gave immunity to smallpox. He printed two letters from local physicians who shared his view, which he also forwarded to Jenner, giving him space for a full reply. It was a typically collegiate touch that mediated the dispute and encouraged the medical community to open debate in a public forum. Beddoes had great talents as an editor, but from now on he would be too preoccupied to exercise them.

The volume was dominated by Davy's three essays, packaged together under the title 'On Heat, Light and the Combinations of Light', and sprawling over its first two hundred pages. The context of a medical journal emphasised Davy's theory of life, and the chemical underpinning it proposed for the famous Brunonian dictum that plant, animal and human bodies were all, correctly seen, lighted tapers in a constant state of combustion. As Brown's system was gradually marginalised, Davy's theory would come to seem less convincing, and he would later grumble that Beddoes had done him no favours by laying his half-baked juvenilia before the scientific public. Yet despite its precipitous leaps from evidence to conclusion, readily discernible even through the fog of purple prose that Davy had draped around them, the ambition and originality of his debut impressed many, and whetted appetites for what was to come.

Beddoes' other publication of the winter was a long-planned tract on consumption, describing the condition in all its variants and stressing the incomparable benefit to society that an effective treatment would bring. It was partly the definitive description of the condition that he had planned for years, partly a booster for the Institution's imminent experiments, and partly a contribution to preventive medicine in general, as advertised in its title, *Essay on the Causes, Early Signs and Prevention of Pulmonary Consumption, for the Use of Parents and Preceptors*. It restated the call he had made in the *Guide to Self-Preservation* for official statistics of contagion and mortality rates, without which the effectiveness of preventive measures could not be properly assessed, and it drew attention once more to the medical profession's failure to support novel and experimental treatments.

It also included Beddoes' findings about the curious absence of consumption in the butcher's trade, accompanied by a report of parallel researches among the fishwives of Musselburgh on the Firth of Forth carried out by a young doctor recently graduated from Edinburgh named Peter Mark Roget. The fishwives of the Scottish ports were a tight, clannish community, usually the daughters of fishing families, skilled from childhood in bait-gathering, preparing lines, and carrying creels of fish to market. As Beddoes had done with the Bristol butchers, Roget interviewed many of them and found the fishwives 'a shrewd and intelligent set of people', fiercely proud of their malodorous profession and 'less subject to the disease than the generality of poor people in this part of the country'.[66]

Dr Roget was a curious and diffident nineteen-year-old who had recently settled in Clifton, and had tentatively begun to socialise with Beddoes and his circle. His father, a Protestant pastor from Geneva, had died of consumption in 1783, and his mother had lived an itinerant life of genteel poverty with her son ever since. Roget had shown great intellectual abilities from an early age, particularly in languages, and by the age of eight had already taken the first steps in his lifelong quest to classify the English language by compiling a dictionary of comparative meanings. He had entered Edinburgh University at the age of fourteen, but the struggle to win the medical degree that might bring him and his mother financial independence had nearly killed him. The two had spent many lonely years in often desperate financial straits that exacerbated anxieties and nervous conditions in both. Roget had finally gained his degree, but had also picked up an ominous and persistent chest complaint.

Roget's mother had spent 1798 touring him around the spas and resorts of the south of England, and Hotwells was, as for so many others, a last resort. Once there, Roget's interests in preventive medicine, consumption and chemical therapy rapidly drew him into Beddoes' orbit, where he met the extraordinary coterie of young men similarly attracted. But aristocrats and heirs, brilliant poets and physicians and chemists as they might be, their freethinking and iconoclastic society, with its militant stance towards the rest of the medical profession, was hardly what Roget, and especially his mother, had envisaged for a young doctor struggling to find a secure place in the world. It was a particular source of chagrin that the role of laboratory superintendent, which would have suited the young chemist-doctor perfectly, was occupied by an extravagant and rather uncouth young man with none of the university training that Roget had nearly killed himself to acquire. He had arrived among a circle of outsiders, to discover that he was an outsider among them.

* * *

As another bitterly cold winter set in, Beddoes' medical duties multiplied. He took Robert Southey who, like Roget, was exhausted with overwork and worry, under his care: Southey had self-diagnosed his life of sedentary study as the root of his ill-health, and had taken to tramping around the countryside in a huge coat, 'like a dancing bear in hirsute appearance'.[67] He was suffering a strange array of symptoms, including a pattering in the chest that he feared might indicate a terminal heart defect but that Beddoes, who was beginning to conceive hypochondria as a growing occupational disease of the leisured classes, tactfully suggested might be more psychological than physical in origin. He prescribed Southey ether, one of the 'volatiles' that he was preparing for trial at the Institution and favoured by Darwin for nervous conditions as well as chest problems: it was, in Brunonian terms, a strong stimulant that boosted the system like a sharp shot of alcohol, but the transitory drunkenness it imparted could also sedate and ease panic attacks. Southey began to use it heavily, 'sorely against my will, for it is very unpleasant to accustom myself to such a stimulus'.[68]

Though the winter of 1798 was as harsh as any before it and food remained scarce, the spectre of riot and revolution was receding. The threats and rumours of invasion from France were binding the nation together, and the martial configuration of the new Consulate that was emerging from the Directory was making it clear that the French had, as Wordsworth would put it, 'changed a war of self-defence for one of conquest', and 'become oppressors in their turn'.[69] It was no longer credible to claim France as the defender of the rights of man; rather, it was all too obvious that her intention was to reduce the rest of Europe to a hinterland of vassal states. Britain, oppressor of liberty as her current administration might be, now seemed liberty's best guarantor in the long term.

This was a bitter pill for the likes of Beddoes to swallow. British troops were, after all, still dragooning Ireland with a savagery that shocked even their commander, Lord Cornwallis; the massacre of civilians had reached a scale far greater than that of the Terror in Paris. The ban on the Corresponding Societies had driven protest from the streets and atomised the dissenters: the mass public rallies against the Gagging Bills, as their organisers had predicted, were now a thing of the past. With Charles James Fox's anti-war party now boycotting a Parliament in which it had no influence, the vacuum of opposition was beginning to be filled by shadowy organisations going under the names of the United Irishmen and United Britons who threatened, in place of political rallies and debates, armed insurrection and paramilitary violence. In this climate of fear the government was settling scores against many of its more persistent critics: Joseph Johnson, for example, had been convicted of sedition on tenuous grounds and sent to the King's Bench prison, from where he continued to publish regardless.

But the majority of those who opposed the government were now resigned to such persecutions, and had little interest in taking secret oaths of commitment to armed struggle in support of a despotic foreign enemy. Many were deciding, with a heavy heart, that it was time to rejoin the national consensus from which they had stood apart since 1792. The likes of James Keir and Tom Poole were volunteering for their local militias, training and organising in the nation's defence, falling into line with the best grace they could muster in the face of the jibes from loyalists who bragged that, in a time of national crisis, they had kept their principles and their nerve while the reformers had abandoned both. Despite Beddoes' sanguine predictions to Watt, it was not the moment for the Pneumatic Institution to claim support by reminding the public of its democratic credentials. But it was, perhaps, as appropriate a time as any to launch a project that, if successful, would assist the war effort on its own terms by fighting in defence of the nation's health.

By the end of the year the purchase of the house in Dowry Square was completed, and work on the building progressed through the remainder of the winter. Laboratory equipment was moved the short distance down the hill from Hope Square, and a wall knocked out of the new house to accommodate a furnace. James Watt paid a visit to supervise the installation of his apparatus in the workshops, and work on the patients' wards began. Finally, on 21 March 1799, the first day of spring, a notice appeared in the *Bristol Gazette* announcing in bold letters the arrival of a 'New Medical Institution', intended 'for treating diseases, hitherto found incurable, upon a new plan'. At this point it was only ready for outpatients, but its services were available to anyone with consumption, asthma, scrofula, palsy, 'obstinate venereal complaints' and 'any other diseases, which ordinary means have failed to remove'. Treatment would be gratis, and 'none of the methods pursued are to be hazardous or painful'. The clinic would be open between eleven and one o'clock, and treatment would be performed by Thomas Beddoes and Humphry Davy.

It was the moment of truth. Beddoes and Davy were ready to begin the long-anticipated experiment; it only remained to see if the sick were equally prepared to be experimented on.

6

Wild Gas

Patients began to arrive in Dowry Square immediately, and in large numbers. The boarding houses of Hotwells gave up their consumptives, their asthmatics, their paralytics and syphilitics, at a greater rate than Beddoes and Davy could examine and treat them. Soon they had eighty outpatients, as many as they could handle until their residential hospital was ready. But to begin with they held back on their experiments: Beddoes had decided that it would be 'prudent to waive the use of gases for a time, and confine ourselves to the administration of common remedies'. He recognised that 'without conciliating that class from which we were to draw our patients, we could make no progress':[1] these sick and needy patients were offering themselves not to be experimented on but to be cured, and it was necessary to begin by winning their trust. He offered digitalis to some sufferers of palsy or paralysis, and nitric acid to some syphilitics, but for the most part limited himself to tried and tested remedies. This was, as he reminded himself, 'perhaps, the first example, since the origin of civil society, of an extensive scheme of pure scientific medical investigation',[2] and he had already discovered the frustrations of attempting to cajole the world into travelling at his own restless pace.

Outside surgery hours Beddoes and Davy kept up their chemical experiments. Davy, still eager to put natural phosphorescence and electricity to the test, was investigating a pair of polished bamboo canes that he had noticed gave off sparks when struck together, an effect he quickly traced to the unexpected presence of silica in the outer layers of the cane stem. Beddoes continued to experiment with various acids as anti-syphilitics, and both worked systematically through the synthesis of gases. In March, Davy picked up a thread of experimentation he had begun two years previously in Penzance, shortly after devouring Giddy's copy of *Considerations on the Medical Use of Factitious Airs*. He had been curious about the paper by the American physician Samuel Latham Mitchill, included as an appendix to the second edition, which examined the properties of nitrous oxide, a compound first synthesised by Priestley and christened by him 'dephlogisticated nitrous air'. Mitchill thought he had discerned in this substance a fundamental quality: just as oxygen had been demonstrated to contain the principle required for combustion and animal life, so nitrous oxide contained its opposite, the principle of contagion and decay that consumed the Brunonian life force. This new air was, he claimed, the deadliest of poisons, and the source of all putrefaction of living matter: wounds became septic and rotted on exposure to it and, if breathed in pure form, 'there will be sudden extinction of life'.[3] He proposed naming it 'septon' and, writing privately to Beddoes, he celebrated their shared passion for Darwinian parody with verses proclaiming

Grim Septon, arm'd with power to intervene
And disconnect the animal machine.

This unsuspected agent in the atmosphere, according to Mitchill, underlay all degenerative diseases: it allowed scurvy to ravage the blood and leprosy to rot the flesh, and showed

Greedy cancer how he best may thrive
And gorge and feast on human flesh alive.[4]

It was an arresting claim, and one that if true would surely promise an immediate breakthrough for pneumatic therapy; but it was easily disposed of and, indeed, Davy had already done so while still in Penzance. 'The fallacy of this theory', he recalled, 'was soon demonstrated by a few coarse experiments'.[5] He had liberated some of the gas from a piece of zinc placed in diluted nitric acid; he had exposed wounds to it without ill-effect, and breathed a small quantity without dropping dead on the spot. It was an example of how flimsy the corroborative testimony that Beddoes included in his books could be; but its inclusion also lent weight to the argument that all such theories were valuable because it was impossible to tell where empirical testing of them might lead. Davy, returning to nitrous oxide as part of a survey of potential therapeutic airs, began to take a new interest in its qualities.

In doing so, he quickly realised that the 'septon' Mitchill had produced was not one substance but at least two. It contained nitric oxide, known from Priestley as 'nitrous air', which gives off red-brown fumes when exposed to common air and is a toxic irritant when inhaled; but nitrous oxide was a separate substance, less reactive, not noticeably toxic and with properties that had never been fully investigated. Priestley had separated it from its poisonous twin but had not experimented with respiring it. Davy promptly wrote a letter to *Nicholson's Journal*, the unofficial clearing-house for British chemical researches edited by the man whose textbook he had devoured as a teenager, to announce that nitrous oxide 'is respirable when perfectly freed from nitric phosoxyd' (the last name his own coinage, following his theory of oxygen and light). 'Mitchill's theory of contagion is of course completely overturned', he continued, and 'the mistake of the Dutch chemists and Priestley probably arose from their never having obtained it pure'.[6]

In the speed and ruthlessness with which he had established his priority, Davy had emulated Lavoisier rather than Priestley: he had taken the work of a previous researcher, stripped it of fallacious theory and reconfigured it, disposing of any other competition with a public announcement and, within days, claiming a virgin field for his own. The seeds of the long-anticipated chemical revolution, he was already beginning to suspect, might lie within its unturned soil. Within Lavoisier's system the fundamental entity was the element: once the essential elements of matter were identified, the remainder of chemistry was conceived to lie merely in elucidating their compounds and combinations. But the oxides of nitrogen showed clearly that this was too simplistic a programme. Nitrogen and oxygen, both elements, might form the toxic orange mist of nitric oxide; they might form atmospheric air, which according to Davy's theories represented their loosest combination; they might form nitric acid, with which he was tentatively treating Bristol's syphilitics; or they might form nitrous oxide, whose properties still awaited discovery. Whoever could establish what caused these two elements to form such different substances under different conditions would arrive at a truly fundamental description of the world, deeper than either Priestley or Lavoisier had glimpsed.

Davy set himself to isolating nitrous oxide from the gas given off by the zinc reaction, and was able rapidly to establish that it could be inhaled with safety, producing only a mild dizzy sensation that might equally well be attributed to temporary starvation of oxygen, or to impurities in the synthesis. His progress was speeded when he recalled that there was an easier way to collect the gas: Claude Louis Berthollet, an associate of Lavoisier, had shown in 1785 that it was given off when crystals of ammonium nitrate were heated gently to around 350 degrees (if heated too fast, they would explode). Davy now obtained some and, on or around 8 April 1799, he and Beddoes made their first full experiment.[7] They placed the crystals in the alembic of Watt's apparatus, sealed the joints around it with the prescribed mixture of clay, lime and borax, and set heat under them. The crystals melted to liquid; as they maintained the heat, the gas began to collect in the hydraulic bellows, gradually displacing the water from the reservoir. Davy sat beside it, inserted a breathing-tube into the air-holder, pinched his nostrils closed to exclude all atmospheric air, placed the other end of the tube between his lips and began to inhale.

The first thing he noticed was a curious sweet taste, followed by the slight dizziness he had felt before. As he continued to breathe, however, a new feeling announced itself: 'a sensation analogous to gentle pressure on all muscles, attended by a highly pleasurable thrilling in the chest and extremities'.[8] But this heightened vibratory condition, it rapidly became clear, was only the prelude to a far more intense and profound effect. 'The objects around me became dazzling and my hearing more acute', he recalled; his senses seemed to be reconfiguring themselves, relaying strange and unfamiliar messages to his brain. He felt as if he were awaking to a world that existed in parallel to the one where he had spent his life thus far, but of whose existence he had until this moment been unaware. As he struggled to grasp what was happening to him, he was swallowed up in a crescendo of sensations, as if every organ of perception was competing to exercise its new-found freedom to the limit.

He had been seated, head bent downwards into the breathing-tube, accelerating into this unfamiliar world with each inhalation; but as the sensations built towards a ringing climax, 'the sense of muscular power became greater, and at last an irresistible propensity to action was indulged in'.[9] At this point Beddoes recalled Davy rising to his feet and becoming a blur of motion, 'shouting, leaping, running' and emitting expressions of exhilaration like one 'excited by a piece of joyful and unlooked-for news' as he rushed around the laboratory.[10] Davy himself retained only vague recollections of the actions 'various and violent' that followed – and, were it not for the scribbled notes that recalled them for him the following morning, 'I should even have doubted their reality'.[11]

* * *

Even in his most messianic projections Beddoes had never anticipated such a discovery. After years of obstacles and delays during which he had been obliged almost daily to justify the notion of pneumatic medicine, and indeed the very idea of medical experiment, the Pneumatic Institution had been open less than a month before his wildest hyperbole had been surpassed. Here was a factitious air, never witnessed in nature, with properties that expanded the scope of chemical medicine beyond anything he had imagined. Nitrous oxide was, if anything, the opposite of the agent of contagion and decay proposed by Mitchill: it seemed, rather, to be a source of excitation and life force more potent even than oxygen itself. It effervesced through the body just as carbon dioxide did through water, flooding the metabolism with an exhilaration of unnatural intensity. It might be regarded as a Brunonian stimulant, in the same category as alcohol but far more potent, except that it seemed to excite the metabolism without exhausting it, as Brown's theory would demand. After his inhalation Davy had slept lightly, ideas fizzing through his head too intensely for profound rest; the next morning he felt no hangover or lassitude, and his eyes were not bleary but gleaming, as one who had seen strange and wonderful things.

There was no doubting the efficacy of the gas, but there were a thousand unanswered questions about how its effects should be construed. By following his Baconian path of untrammelled experiment, Beddoes had entered a *terra incognita* beyond the reach of theory, and it was necessary to explore the terrain more thoroughly before reaching for a tidy explanation. His first communications were tentative. He wrote within days to Darwin of 'something extraordinary made out' at the Institution, but added that it would take time to establish precisely what it meant.[12] On 23 April he informed Jos Wedgwood that he and Davy had made out 'a species of air' that produces excitation 'in the most remarkable manner'; he was unable to resist expressing the hope that, with the new gas, 'we could now stimulate Tom's torpid machine'. His hypotheses were vague – 'it seems to act by giving excitability or life' – but he was in no doubt that the discovery represented a major breakthrough. 'I think', he told Jos with quiet understatement, 'it will realise the expectations and conjectures I originally started'.[13]

He was, however, in no hurry to breathe the gas himself. His oxygen trials had amply demonstrated that he was not frightened of self-experiment *per se*, but in this case he was 'rendered timid' by his 'apoplectic make'.[14] Unlike his assistant, he was no longer young and fit; the spectacle of Davy thrashing and roaring in a strange borderland between frenzy and unconsciousness had left him worried that, with his wheezing chest and heart already overloaded by his bulk, his system might explode under the strain. But the potency of these effects made him curious to try small doses of the gas on some of the Institution's patients. Davy had clearly experienced a surge of muscular pressure and a 'propensity to action', precisely the abilities that palsy sufferers lacked. What might this intense stimulus achieve in those who were unable to move at all?

Beddoes began by selecting a twenty-six-year-old man who, 'after a course of excessive debauchery, especially with regard to fermented liquors', had lost the power of movement down one side. Although he was palsied, there was no 'visible organic lesion' to account for his frozen condition, suggesting that his physical apparatus might be able to respond if stimulated with sufficient force. Davy and Beddoes were obliged initially to move his head down to the mouth of the breathing-tube themselves, but after several treatments the man found he could flex his paralysed arm, and 'at last he could grasp things without a tremor'. Encouraged, they offered the treatment to another even more severely palsied patient, 'as shattered a human creature as can easily be met with'. The gas demonstrated its miraculous powers once more, and the patient soon found himself able to walk without crutches for the first time in many years. Beddoes, after a lifetime of administering foul-tasting medicines and painful courses of treatment, was particularly struck by 'the eagerness with which these patients looked forward for their dose of air, and the pleasure it gave them'. As with Davy, it seemed as if the sensations produced by the gas were reason enough to inhale it, and its curative powers a delightful bonus.[15] It was a medicine whose only side effect appeared to be pleasure.

Administering the gas to patients, Beddoes recognised that it could be given in mild and gently titrated doses, and he gained the confidence to make a trial on himself. As ever, any theory of its action would need to begin with a full description of the condition it produced; but such a description proved no easier for him than it had for Davy. Even for one with such formidable verbal powers, he could only summarise its effects as 'agreeable beyond his conception or belief'. A small dose far exceeded anything he had felt during his trials of oxygen, eliciting sensations of profound pleasure that seemed to have no cause or object beyond the gas itself. Beddoes' prose was stretched to its limits to accommodate this conundrum: he felt himself, he recorded, 'bathed all over with a bucket of good humour', which receded as rapidly as it had arrived, and left 'not one languid, low, crapulary feeling afterwards'. Its warm glow radiated through his lungs and suffused both mind and body. It left him with an appetite healthier than normal that 'led him one day to take an inconvenient portion of food'; but for this discomfort it proved an equally sovereign remedy, removing the unpleasant sensations of distention and indigestion. He never attempted to ride the crescendo of intoxication to the 'higher orgasm' that Davy had reached, but even at lower doses he felt a torrent of strange thoughts rushing through his mind. When acquaintances would say to him, 'Why your eyes twinkle as if you were drunk', he would acknowledge the comparison: he would notice 'unsteadiness' and 'a random feeling' for some hours after inhaling. But the destructive effects of alcoholic intoxication seemed entirely absent, and he confirmed Davy's finding that there was no price to pay, no aftereffects but 'simple high spirits'. It had, in fact, improved his overall sense of health: 'till he took this air', he wrote of himself, in the third person of scientific convention, 'going to a play always brought on a head-ache next morning', but 'now he rises just as fresh as on other occasions'.[16]

* * *

Davy, meanwhile, had taken to self-experimenting regularly. There was much to discover – the volume of nitrous oxide, for example, that was absorbed over a series of inhalations, and the proportion of this that made its way into the bloodstream – but he was also finding it a fascinating recreation, and breathing it simply 'for the sake of enjoyment'.[17] It was by following their self-experimental hunches that they had come upon their discovery in the first place, and now Davy and Beddoes began to expand their trials. The effects of the gas were not restricted to the sick, and indeed Beddoes' self-experiments had suggested that they might equally benefit the healthy. The social circle around the Institution constituted an ideal cadre of volunteer subjects whose responses might suggest new avenues of research. Most of all, though, they were bursting to share their news, and to astonish their friends with the new experience they had to offer.

The first of the circle to try the gas was Robert Southey, who had from the beginning been in the habit of dropping by the Institution whenever he came in to Clifton, being almost certain to catch Davy either at work or in residence. Davy had, in fact, allowed Southey to breathe a little of the impure gas during the first phase of experiments: he had felt a slight dizziness but no more. Now, he offered Southey the improved gas, and an improved method of administration. He had decided that sitting hunched in front of the apparatus, with his head almost between his knees, produced a rush of blood to the head that was responsible for much of the dizziness, and had taken to siphoning the gas from the reservoir into a green oiled-silk bag, as recommended by Watt, its inside dusted with charcoal to remove impurities. He wrapped the breathing-tube tight around the air-bag's open end, filled it, and instructed Southey to hold his nose with one hand, while he himself held the bag and offered it up to his lips.

Southey's reaction did not disappoint. He felt the initial dizziness that he remembered, a moment of vertigo as he sucked the last of the gas into his lungs; then, as Davy removed it, he let out a profound and involuntary laugh, accompanied by a thrill that diffused right through to his toes and fingertips, 'a sensation perfectly new and delightful'. As it ebbed, it left him feeling refreshed and renewed, and for the rest of the day he 'imagined that my taste and hearing were more than commonly quick'. In summary he offered the epigram that 'the atmosphere of the highest of all possible heavens must be composed of this gas'.[18] Away from the laboratory he was more exuberant still. 'O Tom, such a gas has Davy discovered, the gaseous oxyd', he wrote immediately to his brother. 'O Tom, I have had some, it made me tingle in every toe and finger-tip! Davy has actually invented a new pleasure for which language has no name. O Tom, I am going for more this evening, it makes one strong and happy, so gloriously happy! O! excellent air-bag!'[19]

Southey's endorsement was as emphatic as he could make it, and he would become a powerful recruiting sergeant for the green silk bag. Davy was now acquiring ammonium nitrate in quantity, and would soon begin making it himself, following a simple recipe passed on by James Watt that involved boiling woollen rags to release ammonia and capturing the fumes in an acid solution. But his immediate scientific priority was an exhaustive series of animal experiments to confirm the safety of the gas for human consumption and to elucidate its effects on respiration, the changes in metabolism that it produced and the proportion of it that was absorbed into the blood. He began this work with mice and guinea pigs, and would later extend it to include rabbits and kittens, fish and insects. It was not long before he had established that the effects of the gas were largely the same in all animals, and in line with those he had experienced himself. Animals placed in jars filled with nitrous oxide were initially excited: their breathing and heart rate increased as they experienced the crescendo of the intoxication, which was often accompanied by muscle-twitching and convulsions. Gradually, however, the gas would begin to suffocate them: their breathing would slow, convulsions and twitches would pass, and they would become limp and unconscious. If they were returned to atmospheric air at this point, they would recover; if they were not, they would die. As Beddoes had predicted, it was the smaller animals that suffered the most, and the most rapidly.

None of this could be established without destruction, and the animal experiments would remain a gruesome backdrop to all that followed. Findings needed to be confirmed by repetition, and Davy worked his way through a conscript army of flailing, gasping and convulsing subjects. Those that died would be dissected; those that survived would be bled and, more often than not, returned to the experimental queue. In the course of this, Davy was learning much about the parameters of safe dosage for humans; but he was also closing in on some essential truths about respiration itself. Animals, he was discovering, 'are capable of living for a very great length of time in nitrous oxide mingled with very minute quantities of oxygen or common air',[20] just as candles were able to burn it in. Compounds of nitrogen had long been known as 'azotes', incapable of supporting life; but now, it seemed, one of them might play an unsuspected role in the invisible transactions of the breath.

As the evenings lengthened towards summer, it was not merely the sick and suffering who were arriving at the Institution, but the healthy and curious. Robert Southey returned regularly after working hours to drink of 'the wonder-working air of delight',[21] and many others came to experience the new sensation. A loose experimental procedure was set up, with subjects asked to submit brief written observations and descriptions of their experience. Davy, aware that his subjects might have had their expectations raised by second-hand accounts, devised a novel method of randomised control: some volunteers were first given a bag of common air to check that their response was due to the gas rather than their imagination. Dr Robert Kinglake, a local physician and friend of Beddoes, was an early subject. Noting 'additional freedom and power of respiration, succeeded by an almost delirious, but highly pleasurable sensation in the head, which became universal',[22] he promptly offered his services to the project, assisting Beddoes and Davy in supervising the administration of the gases and monitoring the subjects' vital signs. Roget, eager for professional experience, also volunteered as a medical assistant.

The human experiments, as they progressed, revealed an extraordinary spectrum of responses, including several that could hardly have been guessed from the first volunteers. James Webb Tobin, a local physician and friend of the Wedgwoods, inhaling two bags one after the other, seemed to be transformed into another person altogether. 'When the bags were exhausted and taken from me', he recorded, 'I continued breathing with the same violence, then suddenly starting from the chair and vociferating with pleasure I made towards those that were present, as I wished they should participate in my feelings'. At this point Davy entered the room, and Tobin 'made towards him, and gave him several blows, but more in the spirit of good humour than of anger'. His muscular urges then took over entirely, and 'I ran through different rooms in the house, and at last returned to the laboratory somewhat more composed'.[23] It was as if the gas itself contained a hidden personality, with the power to possess its subjects independently of their volition.

These extreme experiences could be disturbing to witness, but to their subjects they were profoundly pleasurable, even compulsive. Stephen Hammick, a surgeon from the Royal Hospital at Plymouth who had been working with Beddoes on the nitric acid treatment, took a small dose which produced only 'yawning and languor'; but a second bag stimulated the familiar 'glow, unrestrainable tendency to muscular action, high spirits and more vivid ideas'.[24] When Davy attempted to ease the air-bag from his lips, Hammick, the effects now starting to grip him power-fully, 'refused to let him have it, and said eagerly, "Let me breathe it again, it is highly pleasant! It is the strongest stimulant I ever felt!" '[25] Davy's animal experi-ments had elucidated the effects of the gas on respiration and muscular action, but they gave no clue to the subjective experience; the human trials, however, revealed an intense mental excitation that manifested itself in an explosion of thoughts and ideas. Unlike the intoxication of alcohol, this state could not be dismissed simply as a dulling of the senses and a slurring of the tongue; rather, it seemed to both subjects and observers that the senses and the imagination were being raised to a pitch of intensity higher than normal life could generate, and were awakening a dormant level of consciousness that was greedy to explore its new reality.

* * *

Throughout the summer, as the experiments continued, the tenor of life at the Institution underwent a marked transformation. As Joseph Cottle later recalled, the nitrous oxide researches 'quite exorcised philosophical gravity, and converted the laboratory into the region of hilarity and relaxation'.[26] By day it remained a clinic for the sick, and a laboratory for Davy's animal experiments; but at day's close it became a philosophical theatre in which the boundaries between experimenter and subject, spectator and performer were blurred to fascinating effect, and the experiment took on a life of its own. Tobin, whose brother was a dramatist, noted an irresistible impulse to throw himself into 'several theatrical attitudes';[27] Richard Edgeworth 'capered about the room without having any power to restrain himself'.[28] This type of bizarre, 'antic' behaviour gradually became an unspoken expectation: subjects happily aban-doned their inhibitions, and vied to produce written reports that encapsulated the experience in memorable aphorisms and turns of phrase. The only price of admission to the show was a brief but very public loss of dignity, and a willing-ness to have a burst of uninhibited speech or action exposed to scrutiny and interpretation. It was a price that not all were prepared to pay: the nervous and censorious Cottle was among those who witnessed the 'laughable and diversi-fied effects produced by this new gas on different individuals' but stood outside the fray, politely refusing the green bag.[29] Those who took the plunge became a brotherhood, bonded by their shared initiation. The wild gas had plainly broken loose, and more literally than Edmund Burke could have imagined.

But for the experimenters, the antics and laughter were only the outward and visible signs of a physical and mental transformation that was tantalisingly difficult to pin down. Its signature effect, a wild burst of laughter with no apparent cause, was a case in point. When the gas was inhaled, the mind seemed to stretch and expand, thoughts and ideas playing across an infinite space before, like the green bag itself, collapsing inwards and deflating. It was at this moment, when the first glimmers of self-consciousness returned, that the laughter, and the shouted phrases simultaneously serious and absurd, tended to bubble up. Although the subject was often unable to explain what was funny, the laughter was nevertheless more than a physical reflex: it was a response to this confusing moment of return, when the embarrassed subject suddenly perceived the gulf between his moment of epiphany and the dazed and gaping spectacle he presented to observers. It was perhaps this same moment of confusion, embarrassment and pent-up energy to which some subjects responded by running, leaping and shouting, particularly once the idea of doing so had been planted by others. Beddoes had been captivated by Galvani's demonstration of the power of electricity to twitch the muscles of a dead frog, proof positive that the animal machine could be directed by man-made stimuli; but nitrous oxide was stimulating not merely gross anatomical functions, but the most exalted reaches of human thought and imagination.

While many of the experimenters followed Southey in seeking to capture the subjective sense of extravagance and pleasure, others attempted to drill deeper into the mental mechanisms that might underlie it. Roget, approaching the gas in the sober spirit of science, felt the familiar sense of vertigo and agitation as the effect approached its crescendo, at which point he noted 'my ideas succeeded one another with extreme rapidity' and 'thoughts rushed like a torrent through my mind'. His metaphor for the cascade of ideas was carefully weighed: it was, he postulated, 'as if their velocity had been suddenly accelerated by the bursting of a barrier which had before retained them in their natural and equable course'.[30] It was a metaphor that drew on the theories of David Hartley, imagining the Newtonian action of nerve-impressions freed from their habitual restraints to produce a bouncing chaos of novelty; but it still begged the deeper question of how this profound derangement of the nervous fibres could be generated simply by a lungful of factitious air.

The difficulty of describing the effects of the gas was also plaguing Beddoes and Davy on the occasions when they administered it to their patients during surgery hours. 'How do you feel?' was a standard physician's question – the fuller the description it elicited, the better the chances of an accurate diagnosis – but

Beddoes was finding that, after a dose of nitrous oxide, patients were frequently at a loss for words. In response to 'How do you feel?', one had typically replied 'I do not know how, but very queer'. There were others, though, who were more ingenious: one had replied, 'I feel like the sound of a harp'.[31] It was a fine description of the way that sounds tended to blur into ringing harmonics as the gas took hold, and it seemed to suggest as eloquently as the poets had done that language itself needed somehow to be hijacked to give an adequate sense of the experience. As one of the volunteers, James Thompson, put it, 'We must either invent new terms to express these new and peculiar sensations, or attach new ideas to old ones, before we can communicate intelligibly with each other on the operations of this extraordinary gas'.[32] Like Lavoisier's chemistry, these were revelations of a magnitude that required a new vocabulary.

It was clear that the human experiments, with their irresistible drift towards introspection, required a 'language of feeling', as Davy called it, in which these subjective states could be properly described.[33] The search for such a language led towards the territory that had been painstakingly explored by their funder, Tom Wedgwood, in the years since he had been forced to abandon the demands of laboratory chemistry for the less physically exhausting, but no less exacting, experiments in the inner laboratory of his sensorium. Tom's notebooks were filled with his attempts to catch the subtle workings of the mind that lay concealed from conscious awareness, and to unpick his sense-impressions to reveal the architecture that generated them. He had spent countless days and nights in such thought-experiments: imagining himself blind and trying to grasp how sight might be extrapolated from touch, or attempting to develop the skills by which a helmsman might train his eye to extract detail from a smudge on the horizon.[34] Was there a similar form of mental exercise that might allow subjects to grasp the hidden mechanisms behind nitrous oxide's explosive effects on consciousness?

Davy, concerned that Tom's enthusiasm and imagination might run away with him, first offered him atmospheric air, from which Tom felt no effect, and which, he pronounced, 'confirmed him in his disbelief of the power of the gas' that sounded too strange to be true.[35] When given nitrous oxide, however, he felt the effects immediately, and 'became as it were entranced', breathing deeply from the air-bag until he had emptied it. At this point he cast the bag aside but 'kept on breathing furiously with an open mouth and holding my nose with my left hand, having no power to take it away though aware of the ridiculousness of my situation'. He distinctly felt his consciousness split in two: one part of him remained in the room, able to hear everything going on around him though unable to respond, while the other was overwhelmed by the sensation of the gas coursing through him 'as if all the muscles in my body were put into a violent vibratory motion', and mischievously prompting him with 'a very strong incli-nation to make odd antic motions with my hands and feet'. He felt lighter than air, 'as if I was going to mount to the top of the room'; he had begun the exper-iment exhausted from a long carriage ride, but the gas wiped away his fatigue, and left him humming with vigour.[36]

It was Tom to whom Beddoes' thoughts had turned as soon as he discovered the gas, and it seemed, just as he had hoped, to supply power and vigour to his 'torpid machine'. But if this first trial raised the delicately unspoken hope that the measure of the Institution's success might be the healing of its patron, Tom's subsequent experiences began to highlight unforeseen dangers in using the gas to treat the sick. Further doses not only failed to achieve the same high spirits, but produced 'rather unpleasant feelings' and disagreeable after-effects.[37] The thrilling vibrations coarsened into dull, pounding headaches; the delightful, floating languors turned to feeble lassitude; the effervescence that the gas left behind it turned to pinpricks and numbness. Tom's constitution, it seemed, was too fragile to withstand the strain of repeated treatments with the gas. As the experiments progressed, Tom would resume a position all too familiar to him: reduced to watching the pleasure of his friends while he lay propped up on a daybed in the corner.

Despite the atmosphere of antic hilarity that set the tone for the majority, Tom was not the only subject to experience disagreeable sensations from the gas. Some subjects found the experience physically uncomfortable from the first, and others developed aftereffects that were considerably more serious. The most significant factor, as the number of experiments increased, seemed to be the gender of the subject. This had not been immediately apparent: Anna, for example, one of the first women to try the gas, had a delightful expe-rience, with 'uniform pleasurable sensations', during which she 'seemed to be ascending like a balloon' and after which she found herself walking the steep alleys back up to Clifton with an unaccustomed spring in her step.[38] But one early experiment caused Beddoes to check his researches with the female sex: a woman, identified in his notes as Miss —N, who had a history of hysteria and cramps, and whom the gas shocked into 'a succession of hysterical fits of consid-erable violence'. Although the effects customarily wore off within minutes, her fits lasted over an hour, and indeed returned over the following days. Beddoes dosed her with opium, the sovereign remedy against overexcitation, but the fits nevertheless returned, and persisted for several more weeks.

Miss —N was, Beddoes acknowledged with regret, 'doomed to be the martyr of this course of experiments'. He was convinced that her response was not simply psychological: she had absorbed a benevolent view of the gas from others, and stated later that she had had 'no idea that air could act so banefully'. But once it was recognised that the gas could, with no discernible warning, produce an experience as intensely negative as the majority were positive, Beddoes resolved to proceed more carefully. Other women were given smaller doses, but still manifested mild hysteric attacks as the effects of the gas reached their peak. Another subject, a Miss R., fell into a curious languor or trance in which 'the power of speaking was suspended, but she was sensible of everything that passed around her'.[39]

The severity of the effect on some of the female subjects exaggerated a gender bias that had been inherent in the circle from the beginning, as it was in all scientific endeavours of the time. Beddoes had been characteristically anxious to correct this inequality, as he had when he gave his series of anatomy lectures to the women of Bristol the previous year; but as the experiments progressed, the volunteer reports collected by Davy became almost exclusively male. The problem lay, perhaps, not so much in female physiology as in the setting in which the gas was administered, and the level of public disinhibition that it demanded. Joseph Cottle records that one female subject, after breathing the gas, became a 'temporary maniac', dashing out of the laboratory and into the street, where she 'leaped over a great dog in her way'. Her performance 'produced great merriment, and so intimidated the ladies, that not one, after this time, could be prevailed upon to look upon the green bag, or hear of nitrous oxide, without horror!'[40] It was the kind of exhibition that could be laughed off or ascribed to scientific experiment far more easily by a man than a woman, however much the experimenters sought to level the field.

*   *   *

Throughout the experiments Davy was a constant and commanding presence, a master of ceremonies who set the tone in ways not always recognised by his subjects, or indeed himself. When a new volunteer arrived to make a trial of the gas, it was Davy who took charge of the laboratory, filling the alembic with the crystals of ammonium nitrate, setting the flame beneath it and presenting the green bag to the latest subject; it was he who delivered the lecture on the chemistry of the gas, the recommended procedure for inhaling it and the expectation of an exuberant burst of pleasurable sensations. Despite his cautious practice of blind testing, he nevertheless created an atmosphere in which antic impulses could be indulged, and after which science and poetry competed to encapsulate the experience. Often, too, he would join in: during the summer of 1799 he recorded that 'a desire to breathe the gas is always awakened in me by the sight of a person breathing, or even by that of an air-bag or an air-holder'.[41] And though he delighted in sharing the experience with others, he took it as frequently in private, dreaming and drifting among the tide of ideas that flowed through him. 'I have often felt very great pleasure when breathing it alone', he wrote, 'in darkness and silence, accompanied only by ideal existence'.[42]

The green bag, which could be pinched tight once it was filled, also meant that he could indulge these solitary experiences outside the laboratory. Over the summer, particularly on evenings when the moon was full, he took to walking along the Avon, past Hotwells Parade, until he was surrounded by the spectacular crags and cascades of the gorge 'rendered exquisitely beautiful by bright moonshine', where he would look down on the hot springs snaking into the river bed at low tide while sipping away at the six quarts of gas that the bag could hold. He often lost consciousness, and once he was obliged to avert a concerned intervention by making 'a bystander acquainted with the pleasure I experienced by laughing and stamping'.[43] He would frequently attempt to capture in his notebook the procession of thoughts that raced through his mind, and the heroic glow with which they filled him. 'I said to myself I was born to benefit the world by my great talents etc.', he recorded on one occasion;[44] on another, that 'I seemed to be a sublime being, newly created and superior to other mortals'.[45] These were ecstatic reveries similar to those in which he had lost himself in Penzance while staring across Mounts Bay at the Atlantic swell, when he had dreamed of taking his place among the exalted sons of genius; but now, when he returned from reverie, he could reflect with equal satisfaction on how fast the gulf between fantasy and reality was closing.

As the summer progressed, Beddoes and Davy began to share their speculations about the medical ramifications of 'the power of the gas, to increase the sensibility or nervous power, beyond any other agent'.[46] Nitrous oxide clearly provoked a fundamental response in the human and animal system, and in particular in the sensations, perhaps the very seat itself, of pleasure and pain. The experiments had repeatedly demonstrated overwhelming sensations of pleasure; in many ways, the problem of describing the effects of the gas was the problem of describing pleasure itself. It had been characterised in many ways: by Beddoes himself as like receiving a piece of unexpected and joyful news, by James Webb Tobin as the feeling produced 'by reading a sublime passage in poetry when circumstances contribute to awaken the finest sympathies of the soul'.[47] Beneath these metaphors, however, must lie a palpable physiological response: under the influence of the gas pleasure arose without any cause beyond itself, just as laughter rose to plug the gap that language was unable to fill.

But if nitrous oxide directly stimulated the seat of pleasure, then it followed that such a seat must exist, somewhere in the body or brain or perhaps diffused through both; and, furthermore, that it could be stimulated by purely chemical means. Both of these were extraordinary discoveries, and 'the Pneumatic Institution', Beddoes declared, 'might advance a fair claim to the premium, anciently proposed for the discovery of a new pleasure'. The discovery of new pleasures had been a hallmark of the eighteenth century, but its final year had now produced the most novel of them all: one that pointed to a future where man might 'come to rule over the causes of pain or pleasure, with a dominion as absolute as that which at present he exercises over domestic animals and the other instruments of his convenience'.[48] It was a turn of events to rival any of the extravagant spoofs of *The Golden Age*, with its purple cows and hedgerows dripping with butter. Within weeks of its establishment, the Institution had achieved the direct synthesis of pleasure itself, which was now available on demand from an air-bag ('Things strange to tell! Incredible, but true!'). Yet there were already signs that it would prove difficult to convey the enormity of the discovery to the outside world. Even Davies Giddy, hearing the accounts of the Institution second-hand from Cornwall, worried that the partners he had introduced were bringing out the worst flaws in one another. 'I entertain the highest opinion of Beddoes and Davy', he confided to Jos Wedgwood, 'but it is impossible not to wish that they would have more frequent recurrence to the dictates of their reason and indulge less frequently in wild flights of unrestrained imagination'.[49]

But pleasure was only half of the equation: just as the gas afforded 'the most delicious of luxuries', it must also, once properly understood, offer 'the most salutary of remedies'.[50] Davy was beginning to suspect that the gas operated on nerves that generated both pleasure and pain, and that a substance that increased one must by the same token diminish the other. He had already begun some tentative experiments along these lines: 'having accidentally cut one of my fingers so as to lay bare a little muscular fibre', he had stuck the bleeding finger into a bottle filled with nitrous oxide. The dripping blood turned a livid purple, and the pain 'was neither alleviated nor increased'; but when he removed the finger, 'the wound smarted more than it had done before'.[51] He had also noticed that, at the peak of intoxication from the gas, the pain of headaches and toothaches was 'for a few minutes swallowed up in pleasure',[52] and he was now beginning to work his observations into a physiological theory. 'Sensible pain is not perceived after the powerful action of nitrous oxide', he proposed in his notebook, 'because it produces for a time a momentary condition of other parts of the nerve connected with pleasure'.[53]

Cutting his finger down to the muscle was the type of accident to which Davy was prone. His mind raced ahead of his physical capabilities, and despite his experimental genius he could be quite slipshod in the peripheral business of rigging his equipment and taking elementary safety precautions. He regarded cuts and burns as trivial distractions to be ignored. From July onwards, however, his self-experiments began to take a more deliberate, dangerous and sometimes alarming turn. If any of the other factitious airs held similar surprises to nitrous oxide, he wished to discover them for himself, and before any of his competitors. Some of these airs were held to be extremely toxic; but so had nitrous oxide been, and if he had been happy to accept the received wisdom of others he would never have experimented with it in the first place. He resolved to make trials with every air that he could synthesise until he was confident he had the measure of them.

Some defeated him at the first hurdle. Carbon dioxide produced such agonising spasms in the throat and lungs that he was unable to inhale it, and only when he diluted it with air was he able to establish that, beyond its noxious qualities, it produced no sensation beyond giddiness and a slight sedation. When nitric oxide was inhaled, 'nitric acid was instantly formed in my mouth, which burned the tongue and palate, injured the teeth, and produced an inflammation of the mucous membrane which lasted for some hours'.[54] Other airs he was able to inhale more deeply. Hydrocarbonate – hydrogen mixed with a large proportion of carbon dioxide – proved easily respirable, though he was aware from his animal experiments that it was fatal even after relatively brief exposure. After several inhalations in succession he had 'just enough power to drop the mouthpiece from my unclosed lips' before he found himself 'sinking into annihilation'; he returned to consciousness gradually, his ability to distinguish shapes and objects returning only gradually as he gasped for air. 'On recollecting myself', he wrote, 'I faintly articulated, "I do not think I shall die" '.[55] He inhaled nitrous oxide and oxygen to bring himself round, but they failed to produce their customary exhilaration, and he woke the next morning still weak and feeble – and well aware that, had the mouthpiece lodged between his lips before he passed out, he would not have woken at all.

Such experiments could be disturbing for visitors to witness: as Joseph Cottle recalled, Davy 'seemed to act, as if in the case of sacrificing one life, he had two or three others in reserve on which he could fall back in case of necessity . . . I half despaired of seeing him alive the next morning'.[56] But Davy was an ambitious young researcher in a competitive field: those who were prepared to take such risks would travel faster and further than those who were not, and the unpleasant and dangerous experiments were as necessary as the pleasurable ones. If nitrous oxide had such a powerful excitative effect on the metabolism, then other airs should, according to the Brunonian scheme, have equally powerful depressive ones. 'I have been minute in this account of the experiment', Davy wrote of the trial that he had pushed almost to the point of death, 'because it proves that hydrocarbonate acts as a sedative'.[57] Yet the determination with which he applied himself to these trials, and the cool objectivity with which he conveyed the results, suggests an element of self-initiatory ordeal, a deliberate search for his own limits. He was defining himself, and his genius, by his willingness to push experimentation to the utmost, taking on the roles of pioneering researcher and fearless subject simultaneously. The part in which he had cast himself was a heroic one, and it demanded heroic gestures to mark him out.

*　　*　　*

By the late summer, as word of the nitrous oxide experiments began to make its way into the wider world, a curious new property of the gas emerged: it seemed to lose potency when experiments were attempted outside Bristol. Joseph Priestley's son, Joseph junior, participated in the antics at the Institution, and wrote excitedly to his father about them; but the original discoverer of the gas wrote back from Pennsylvania that his own trials had produced no more than an unpleasant throbbing in the head that made him discontinue after a few inhalations. James Watt and Matthew Boulton in Birmingham, as soon as they heard of Beddoes and Davy's discovery, had promptly manufactured their own supply of nitrous oxide, but were 'much less affected' by it than the Bristol circle, leaving them wondering if they had synthesised the wrong gas by mistake.[58]

But another explanation for this curious phenomenon was suggested by Maria Edgeworth, who came to Clifton in the early summer and was sufficiently intrigued by the extraordinary tales emanating from the Institution to want to observe the antic experiments. 'A young man, a Mr. Davy, at Dr. Beddoes'', she wrote drily to a friend, 'enthusiastically expects wonders will be performed by the use of certain gases, which inebriate in the most delightful manner'. It was clear to Maria from the testimony of the experimenters that 'pleasure even to madness is the consequence of this draft', but 'faith, great faith, is I believe necessary to produce any effect upon the drinkers'. Though many volunteers were carried away by the experience, there were some for whom the anticipated effects simply failed to materialise, 'adventurous philosophers who sought in vain for satisfaction in the bag of gaseous oxyd, and found nothing but a sick stomach and a giddy head'.[59] There must be more to the experience than simply the chemical properties of the gas: after all, where were the comparable few who could drain a bottle of brandy without feeling drunk?

From the beginning both Beddoes and Davy had recorded trials where the subject had felt little effect from the gas, but neither had dwelt on them beyond speculating that the dose was probably insufficient: given so many extravagant and glowing testimonies to nitrous oxide's power, the damp squibs had received little attention. Within the Brunonian paradigm it was well understood that some subjects were more constitutionally sensitive to stimulants than others. But, as the experiments slowly dissipated with the summer, the failures began to mount, even among those on whom the gas had previously worked. It was not only Tom Wedgwood who felt a powerful and pleasurable effect at his first attempt but disappointment and disagreeable feelings thereafter. Robert Southey, returning to the fray in the autumn after a recurrence of his illness had kept him away, found that the atmosphere of heaven had soured. His sensitivity to the gas had grown – 'half the quantity affects me, and its operation is more violent' – and the sensations it produced were no longer 'in the slightest degree pleasurable'.[60]

Davy and Beddoes had done plenty to establish that the effects of the gas were no mass hallucination. Not only did they have their own self-experiments to rely on, but also Davy's animal experiments and measurements of physiological changes to blood, pulse and breathing. The blind tests in which common air was substituted were invariably detected by the subjects. But there were other elements in play beyond the purely chemical. Davy, even if unconsciously, was creating a framework for the experience, and a language in which it might be understood. His enthusiasm for the gas was contagious, and the subjects who volunteered did so in expectation of intoxication and pleasure. But it is uncertain that anyone else investigating nitrous oxide in 1799 would have pursued these effects in the first place, or explored them with the extravagant abandon that characterised the Institution's researches. Most would, like Priestley, Boulton and Watt, have inhaled a few times, noticed a fullness in the head and a throbbing in the temples, checked their pulse and discontinued their self-experiments. It was the combination of Beddoes and Davy, and their interests not merely in medicine and chemistry, but in the roots of perception and sensation, that tempted them to explore more fully nitrous oxide's curious tugging at the edge of the conscious mind; and it was the presence of a circle of friends who shared their passions that shaped the progress of the subsequent investigations.

Here, perhaps, was one of the paradoxes to which experiment was prone, and which had been hinted at by Joseph Priestley. Truly empirical researches might not only lead in unexpected directions, but might also in the process acquire their own idiosyncratic momentum, and accelerate towards perverse conclusions. The nitrous oxide researches were not based on falsehood, nor were Beddoes or Davy delusional; but the method they developed was nevertheless a solipsistic one, with each further trial confirming and enriching previous findings while, as Maria Edgeworth observed, making them ever more incomprehensible to those outside the laboratory. They had indeed made a momentous discovery, and it had sent them in a direction hitherto unexplored; but it was a direction to which it became ever harder to leave meaningful signposts. Their observations and descriptions of the gas built upon one another, capturing in rare detail an unfamiliar state of mind but creating a closed circle and a self-referential language as they did so. Their self-experimentation gave them, from their perspective, the perfect response to any sceptic – try it and see for yourself – but there were few prepared to make this leap of faith, and even among those who did there were many who found nothing but dizziness and a buzzing head.

It was a problem of which Beddoes was not unaware, and which he would address in his first report on the experiments. He stressed that it was essential to keep firmly in mind the distressing failure of medicine to progress into the modern age, which meant that there could be 'no question about how far it is necessary to strike into a new path'. To pursue novelty is essential: it is by definition risky, but to take risks 'requires only courage; and it is some reproach to the age, that it should require so much'. The real challenge at this point becomes that of finding the most productive path among the plethora of unexplored tracks, which requires not only courage but intelligence, instinct and planning. 'Whether the persons concerned in the present undertaking have been able to discover the right path' was an open question, but Beddoes nevertheless felt they were justified in taking the road that they had.[61] The discoveries they had made were too remarkable to be of interest to them alone, and even if they were not what he or anyone else had expected, the world must sooner or later find a use for them.

It was an eloquent defence of the Institution's programme; but perhaps the element of the experiments that made them so hard to transplant outside Bristol was not scientific but social. The way in which the researches developed was a manifestation not simply of the gas, but also of the joy that the participants took in each other's company. Over the preceding years the world beyond their circle had become almost too cold and dark to bear. The decade had begun with the promise, and for many the firm expectation, of a world remade, freed from tyranny in all its forms; now it was concluding with nothing in prospect but tyranny on all sides. Beddoes, who had brought the circle together, was glad of an excuse for celebration, and for Dowry Square to become a microcosm of a future that, outside its doors, had become no more than a dream; and Davy was the perfect catalyst to produce this bubble of happiness. Only ten years old when the Bastille had fallen, he had never invested in the hopes that the rest of the circle had nurtured, nor been shattered when they had been extinguished. Innocent of their tragedy, he was an avatar of a brighter day that might yet eventually dawn.

*   *   *

As the cold, wet autumn drew in, the antic frenzy of the nitrous oxide researches began to dissipate; and as it ebbed, it revealed the strain under which Davy had put himself. He had spent nearly a year working 'from twelve to fourteen hours a day in laborious experiments', most of the time in a laboratory full of dissected animals and 'the fumes of the evaporable acids or ammonia'.[62] He had usually been too excited, busy or nauseous to eat properly, and had rarely taken the time to wash or change his clothes: it was during this period that he began his lifelong habit of putting fresh socks and shirts on over worn ones. His 'health was rather injured' from inhaling gases; he blamed the toxic airs, such as hydrocarbonate and carbon dioxide, rather than the nitrous oxide, but he had developed a weakness in his chest that would never entirely leave him, and to which some would later attribute his premature death.

With Beddoes' encouragement he took leave to visit his mother in Penzance, where he hoped that home cooking, sea air and rest would restore him (although he also took with him equipment to analyse the chemical composition of cuttlefish ink). When he stepped off the coach at Marazion at the far end of Mount's Bay to pay a visit to his aunt, she was horrified by his pale and drawn appearance; he had arrived ahead of the mail coach, and she insisted that he stay with her until his mother received the letter announcing his arrival, to spare her from being 'doubly alarmed at his unexpected visit and altered looks'.[63] He ate and slept and walked the shores and cliffs; and when he returned to Bristol two weeks later, he went directly to the laboratory and inhaled nine straight quarts of nitrous oxide, a far larger dose than usual. As the gas took hold, he shouted loudly, 'What an amazing concatenation of ideas!'[64] He was sufficiently restored to resume his work – and, indeed, to push his experiments even further.

Meanwhile, Beddoes had rushed into print. His interim report, *Notice of Some Observations Made at the Medical Pneumatic Institution*, emerged at the end of October, published by Joseph Cottle in Bristol and by Thomas Longman in London, a connection made by Coleridge, who had returned from Germany and was now writing short-order journalism in the capital. Although Beddoes had a spectacular success to report, it was not an easy one to communicate. He needed to lay before his readers and subscribers the revolutionary implications of his discovery while also maintaining the caution proper to preliminary findings; he had to broach an entirely novel form of research into the science of consciousness while acknowledging to his subscribers that these were experiments quite different from those to which they believed they had been contributing thus far – and while urging them to contribute further. He needed to make the most of his dramatic early successes with cases of palsy, but also to manage the expectations that he had been stoking for years about the prospects for the treatment of consumption, and to explain that the most remarkable effects of nitrous oxide had, thus far, no clear therapeutic dimension.

His tone, however, was of conviction soundly vindicated. He had been right to suspect that the factitious airs held unsuspected potential, and he had not been discouraged by the 'reptiles that plant themselves on the high road to improvement, and try to hiss back all who would advance'.[65] He had persevered and, having found a superintendent 'equal to my wishes and superior to my hopes', had begun his programme of systematic experiment on all the gases, putting his theories and speculations aside and trusting that 'among the possible medical applications of these agents, some one would occur as particularly worthy of our minutest attention'.[66] His empirical method had been rewarded in spectacular fashion, as nitrous oxide had revealed effects that 'I find it entirely out of my power to paint'.[67] He gives his observations of Davy's first experience, and describes his own trials at length, followed by the confirmation of two dozen further experimental subjects, and he details the responses of his first two palsy treatments. He stresses caution – the gas must be correctly prepared, and administered only in appropriate cases – but announces with confidence that the era of pneumatic medicine is now upon us. These discoveries have already expanded the horizons of scientific knowledge, and point ahead to a future medicine where 'man may, sometime, come to rule over the causes of pain and pleasure'.[68]

This brisk and sanguine summary is followed by a torrent of suggestions for expanding on the Institution's discoveries. A new quarterly journal is to be launched, with the title *Researches concerning Nature and Man*; its first issue will feature 'a paper on the philosophy of medicine by the author of this notice', and 'part of a vast chemical investigation, connected with the gas', by Davy.[69] More physiological research must be done: an exhaustive series of animal experiments to make 'a complete chemical acquaintance with the function of the lungs, the stomach, and the skin'.[70] An expert anatomist must be recruited. The tone is breathless, with perhaps a touch of the mania that accompanied the conception of the project seven years before. A postscript is added, made possible by a delay in publication: a new palsy patient, a female aged fifty-five, has reported after several treatments that 'my head has not been so well these two years'. A second delay permits a second postscript, containing more of the 'important information' that is accruing daily: the patient is recovering her movement, raising her arms and walking without support. Her recovery is exceptional, and should be 'a source of self-congratulation' to the Institution's subscribers.[71]

But despite its news of groundbreaking discoveries, Beddoes' report can hardly have been a reassuring document for those subscribers to receive. The grounds on which he had been selling the Institution so passionately and for so long seem to have been quite forgotten. Its *raison d'être* from its conception was the treatment of consumption: it was on this basis that Erasmus Darwin gave it his imprimatur, and it was only months since Beddoes had issued a tract specifically dedicated to presenting consumption as the greatest challenge faced by medicine. Yet now, he informs his readers, 'I easily gave up my first hypothesis respecting consumption'; the research has led in other directions, and the disease is not mentioned again beyond a vague confidence that 'certain gases or vapours' may still prove beneficial to it.[72] Beddoes is not blind to the anxieties that he may be causing his supporters, and among his blizzard of proposals is one for a steering committee to oversee and approve new expenditure; but he reminds his readers that the Institution was never intended 'for the sake of relieving that distress which arises from poverty, but that which arises from the imperfect state of medicine'.[73] The Institution is more than a hospital: it must be permitted to raise its sights above day-to-day treatment and open up new vistas that hold the possibility to heal not simply the sick, but medicine itself. And to accomplish this, experiment must be allowed to lead where it may.

* * *

At the end of October, Coleridge returned to Bristol for the first time since his year of wandering across Germany and the Continent. He paid a visit to Cottle, who found him juggling a vaster portfolio of projects than ever: political essays spun out from the leading articles he was now writing for Longman's *Morning*

Post; collaborations with Southey, whose banishment he had now rescinded; the fruits of his German studies, including a translation of J.F. Blumenbach's monumental *On the Natural Varieties of Mankind*; more poetry for a second edition of the *Lyrical Ballads*, and various 'money-books' to generate the advances to keep all these other plates spinning. Cottle told him of the Pneumatic Institution and the extraordinary discovery of nitrous oxide, and took him down to Hotwells to meet Davy.

Coleridge and Davy's friendship would evolve and endure through the many phases of their future careers. Initially Davy, as the poet's acolyte, would draw on Coleridge's genius to construct the public persona with which he would achieve greater scientific fame than anyone since Newton; in this guise he would later be instrumental in rescuing his burned-out mentor's reputation and drawing him back into public favour. But all began with a green silk bag of nitrous oxide. As Coleridge inhaled and felt its warmth diffusing through his body, he did not reach for extravagant metaphors, but stated precisely that the sensation resembled 'that which I remember once to have experienced after returning from the snow into a warm room'. He felt no muscular urges beyond a desire to laugh at the company who had assembled to watch him, and the sensations faded almost as soon as they had begun. His second experience, 'after a hearty dinner', was milder; but 'the third time I was more violently acted upon'. This time he 'could not avoid, nor felt any wish to avoid, beating the ground with my feet'; and 'after the mouthpiece was removed, I remained for a few seconds motionless, in great extacy'. He followed this with a fourth trial, which initially had so little effect that he suspected he had been given common air, but he soon felt the telltale 'warmth beginning about my chest and spreading upward and downward', and as the effects reached a peak he found himself in a state 'of more unmingled pleasure than I had ever before experienced'.[74]

For Coleridge, as for the rest of the circle, his impressions of nitrous oxide were intimately mingled with those of the laboratory superintendent, who had an equal ability to breathe energy and excitation into his experimental subjects. 'Every subject in Davy's mind', he later wrote, in terms recalling those that others had used to describe the gas, 'has the principle of vitality. Living thoughts spring up like turf under his feet.'[75] Davy, in turn, found Coleridge possessed of the poetic genius he had been led to expect: among the pieces he heard him recite was the unfinished 'Christabel', which Coleridge had had 'scarce poetic enthusiasm enough to finish' until he saw that 'Davy is so much delighted' with it.[76] Both were now in collaboration with Robert Southey: to Davy's Zoroastrian epic was added Coleridge's new scheme to co-author an eight-volume life of Mohammed in hexameter verse.

But Davy was also quick to appreciate that Coleridge's expansive vision had the power to transform science as fundamentally as poetry. In the German universities he had encountered a new generation of scholars who were outstripping their British counterparts across a huge range of fields, from science to history to philosophy, and where the ideas of Immanuel Kant that he had first explored with Beddoes in Rodney Place were being widely applied and extended. Johann Fichte and, particularly, Friedrich Schelling had taken up the Kantian premise that the world as we perceive it is shaped by the architecture of the human mind, and had begun to map out a new metaphysics that projected this vision onto the contours of the natural world. Schelling's *Naturphilosophie*, as he christened it, announced the death of materialist philosophies and their arid view of the world as an assemblage of lifeless corpuscles: 'the powers of chemical matter', he asserted, extend far 'beyond the boundaries of the merely mechanical'.[77] The future would witness the rise of a new and dynamic vision of nature, in which the solidity that it presents to the human mind would be recognised as an illusion concealing a deeper reality: a flux of vital energies and opposing forces of the spirit from which the appearance of stasis is generated.

Since his return Coleridge had launched into an intensive study of Kant that was in the throes of reconfiguring his entire philosophy. In it, he had found a way of framing and intensifying the role of subjective experience, and limiting the claims of Enlightenment science and its implicit atheism. He was leaving behind him the materialist speculations of Brown and Hartley: as he would later put it, 'Association in philosophy is like the term stimulus in medicine: explaining everything, it explains nothing; and above all, leaves itself unexplained'.[78] He had named his first son Hartley, but his second would be christened Berkeley, after the bishop whose insistence that the world could only truly exist in the mind of God had led him to a radical philosophy of 'immaterialism', or idealism.

The new German idealism offered a world view that allowed Coleridge, and henceforth Davy, to navigate a slender channel between the rocks of French revolution and British reaction, and pursue a route to progress that was tainted by neither divisive ideology. Germany was a powerhouse of research and progress, but its novelties, unlike those of France, were seen as purely intellectual and unconnected with any political programme. The Kantian philosophies were revolutionary, but it was a revolution that pointed the way to inner freedoms and the primacy of the life of the mind in forms that could evolve under the political lockdown of the absolutist Prussian state. By creating space for an ineffable vital force behind the visible world, they enabled science to shake off the taint of godless materialism that had become so damaging to it, and to create a politically neutral language that it could speak without being forced into locking horns with the defenders of tradition.

Under its banner Coleridge was now explicitly distancing himself from the political causes to which he had been so passionately committed in the days of *The Watchman*. He had, in a phrase frequently repeated in his notes and letters, 'snapped my squeaking baby-trumpet of sedition': he now, as he had just announced in a letter to Wordsworth, regarded the French Revolution as 'a complete failure',[79] and political engagement as a snare that entrapped the mind by forcing it into unwinnable conflicts not of its own making. The German idealists had become his lodestar in mapping a trajectory inwards, towards a self-actualisation invisible to the outside world, but which offered new and unsuspected freedoms in its interstices.

Davy, for his part, needed little encouragement to postulate a science of life that was animated by vital and mysterious forces: he had, in his essays on phosoxygen, already located such forces within the action of light and heat. But the version of Kant's philosophy that he was now absorbing from Coleridge underpinned the further idea that the mind that recognised this process was not an impartial observer, but was somehow constructing these forces in its own image. His notebooks from this time record a new subjectivity in his conception of the material world: a moment, for example, sitting on a rocky outcrop in the lee of the Avon Gorge, when 'for the first time in my life, I have a distinct sympathy with nature'. He was no longer a tiny spectator in the theatre of nature's clashing elemental forces, but a player on the stage, where 'everything seemed alive, and myself part of the series of visible impressions; I should have felt pain in tearing a leaf from one of the trees'.[80] In his poetry Davy attempted to capture the sensation that his experiments with the gas had torn the curtain of the material world aside, and dissolved the boundaries between perception and creation. 'On Breathing the Nitrous Oxide', in his notebook of late 1799, portrays in pulsing incantation a surrender to the vital forces of excitation, in which they transformed not only his senses but the reality around him:

Not in the ideal dreams of wild desire
Have I beheld a rapture-wakening form
My bosom burns with no unhallow'd fire
Yet is my cheek with rosy blushes warm

Yet are my eyes with sparkling lustre fill'd
Yet is my mouth replete with murmuring sound
Yet are my limbs with inward transports thrill'd
And clad with new born mightiness around.

The influence of Coleridge on Davy's understanding of nitrous oxide would soon become apparent; but the influence of nitrous oxide on Coleridge remains harder to gauge. His written account for Davy would seem to situate it, for one so fascinated by the mechanics of perception and sensation, as a landmark experience. Yet he would never write about his trials with the gas beyond that note for Davy, nor did he refer to them in any of his surviving letters: his voluminous correspondence with Davy would include only a few references to the gas, and none of them to his own experience of it. Even his private notebooks leave it unmentioned, as if he was reluctant to acknowledge to himself that the episode had taken place. Within the same notebooks are dozens of intensely focused observations and descriptions of the ways in which our minds take an active role in constructing the world around us, where the most mundane sensations are deconstructed in penetrating detail; yet a succession of experiences that climaxed with the greatest ecstasy he had ever felt left, for whatever reason, his monologue uncharacteristically stilled.

This silence could, however, be seen as characteristic in other ways. Coleridge's opium use, for example, which was shortly to spiral into chaotic dependency, would also remain unmentioned until its severity forced it into public view, when it would be confessed only painfully and as a last resort. His acute self-examinations, undertaken to uncover the source of a curious sensation or vision, would often include his diet and digestion, his minor preoccupations and anxieties, while contriving to ignore the small matter of the dozens of drops of laudanum he was consuming nightly. It would be left to his later acolyte, Thomas de Quincey, to examine and confront opium's powers explicitly, and also to lay the unfortunate weakness of his mentor before the literary public in unvarnished form. Coleridge's profound ambivalence about abandonment to the voluptuous, also a hallmark of his romantic infatuations, made unalloyed pleasure a furtive and guilty experience, and perhaps imbued his surrender to the sensual embrace of the gas, in public view, with a retrospective sense of shame.

But there was also, perhaps, another reason: for all the ineffable qualities of the gas, Coleridge could sense the direction in which the researches must ultimately tend. He had arrived late to the party, and the language in which the experience was framed had already been formed. The dimension of sensual pleasure had been explored fully and frankly, and placed at the centre of its effects by Beddoes, for whom, as for Erasmus Darwin, the increase of happiness was a laudable end in itself, and indeed the highest reasonable expectation of life. The poets had sought a more rarefied language that expressed sensations beyond mere pleasure, but they were typically cast in terms of the 'sublime', classically defined by the young Edmund Burke as 'that state of the soul, in which all its motions are suspended'.[81] With its dimensions of awe and a cosmic fracturing of the human sense of scale, 'sublime' was a term that allowed for forces more potent than reason and not susceptible to it, but one that stopped short of imputing them to a divine cause. Southey, for example, when he had used transcendent language such as 'the atmosphere of the highest heaven', had done so in teasing reclamation of the epiphanies of religion, ironic in its recognition that the experience, however exalted its stimulation of the highest faculties, was a factitious and chemical one that had been administered in a laboratory. For Coleridge to acknowledge the experience as truly transcendent would be to bring the faultline between materialism and religion into a focus that he wished to avoid. By trapping transcendence itself within a material cause, it threatened to reduce the religious sense to chemistry; and Coleridge, as William Hazlitt would later observe, 'always somehow contrived to prefer the unknown to the known'.[82]

* * *

Coleridge left Bristol as he had arrived, caught in a tug of war between passion and duty. He was expected back at Nether Stowey by Sara and Hartley, whom he had not seen for well over a year, and he had urgent journalism assignments offering him desperately needed income in London; but he put both aside to pay his first and long-anticipated visit to Wordsworth in his homeland of the Lake District which, he recognised immediately, offered the perfect 'quiet spirit-healing nook' for his inward turn of mind.[83] He left, late and reluctantly, for London, and in December Davy paid a visit to the capital to join him. He was powerfully struck by his first glimpse of the metropolis, and the richness of its overlapping scientific and literary circles. He dined out with the celebrated philosopher William Godwin, and became the sounding board for an extended rehearsal of Coleridge's new political position: he had just completed a long essay on Napoleon's new constitution for the French Republic, and was currently writing a *Morning Post* leader under the banner 'Advice to the Friends of Freedom'. It was time, he argued, for all British citizens to put faction aside and close ranks: 'too much of extravagant hope, too much of rash intolerance, have disgraced all parties'. The rhetoric of liberty had served only to strengthen the hand of despotism: in the times that now lay ahead, 'those who love freedom should use all imaginable caution to love it wisely'.[84]

On his return Davy immediately gulped down seven quarts of nitrous oxide, which produced 'the usual pleasurable effects';[85] but in the last week of the old century, he set himself to push his self-experimentation further than ever before. There were still questions about the effects of the gas that remained unanswered. Because of the brevity of its action, it was hard to gauge whether it functioned as a true Brunonian stimulant: if it did, then an extended administration would be expected to reveal that, like alcohol, its temporary excitation was followed by a noticeable debilitation and depression of the metabolism. On 23 December 1799, Davy put this theory to the test by drinking down a bottle of wine in eight minutes, noting as he did so that alcohol at high dosage produced 'a sense of fullness in the head, and a throbbing of the arteries' not unlike the gas itself.[86] He felt a loss of coordination, the fullness in the head became a thumping headache, and within an hour he had passed out. Waking with a headache and nausea, he breathed both nitrous oxide and oxygen but found that neither produced a perceptible effect. He persisted with the nitrous oxide, continuing until its familiar effects took hold and he 'made strides across the room, and continued for some minutes quite exhilarated', before vomiting and collapsing into a listless stupor.[87] Taking stock the next morning, he decided that the experiment had been inconclusive: the debility he had felt was essentially the effect of the wine, and the gas had neither aggravated nor alleviated it.

Since combining nitrous oxide and alcohol had proved impractical, he decided to approach the problem from the opposite direction: rather than using alcohol to extend the effects of the gas, he would extend his exposure to the gas until the effects lasted for the same length of time as those of alcohol. For this, he realised, he had the perfect apparatus: a breathing-box that had been designed by James Watt for patients who were too ill to hold the breathing-tube to their lips. It was essentially a sedan chair enclosed inside a panelled chest, forming an airtight chamber that could be filled with gas pumped in from the reservoir; the nitrous oxide could be mixed with air or oxygen to render it breathable, and all the gases could be remixed and replenished as necessary. On Boxing Day, Davy enlisted the physician Robert Kinglake as his supervisor for an experiment that would dramatically extend the dosage and duration experienced thus far. He stripped to the waist, placed a thermometer under his armpit and entered the box, instructing Kinglake to release twenty quarts of gas into it every five minutes. Davy sat in the sedan chair absorbing a mixture of gas and air steadily for an hour and a quarter, until he felt the familiar muscular tremors and urge to laugh, and 'luminous points seemed frequently to pass before my eyes'.[88] At this point, hoping that he had saturated his system, he emerged from the box and began to breathe a further twenty quarts of pure nitrous oxide.

At this level of exposure the gas took Davy to a dimension he had not previously visited. Objects became dazzling in their intensity, sounds were amplified into a cacophony that echoed through infinite space, the thrillings in his limbs seemed to effervesce and overflow; but then, suddenly, he 'lost all connection with external things', and entered a self-enveloping interior realm of the senses. Words, images and ideas jumbled together 'in such a manner, as to produce perceptions totally novel': he was no longer in the laboratory, but 'in a world of newly connected and modified ideas', where he could theorise without limits and make new discoveries at will. After what seemed an eternity, he was brought back to earth by the sensation of Dr Kinglake removing the breathing-tube from his mouth; the outside world seeped back into his 'semi-delirious trance' and, as the familiar energy flowed through his limbs, he began to pace around the room. But a part of him was still present in the dimension of mind that had swallowed him whole, and he struggled for the words to capture it. He 'stalked majestically' towards Kinglake 'with the most intense and prophetic manner', and attempted to shape the insight that had possessed him. 'Nothing exists but thoughts!' he blurted. 'The world is composed of impressions, ideas, pleasures and pains!'[89]

This exclamation would stand in Davy's eventual published account as the summation of his nitrous oxide experiments, and over the century to come would enter the pantheon of great epiphanies of scientific discovery. Yet its later presentation as a spontaneous 'eureka' moment, as with many such moments, was more carefully constructed than it appeared. Davy's first experience of the gas, as recorded by Beddoes, was a stomping, bellowing chaos from which no coherent insights could be recovered, and which indeed could barely be remembered. Davy's own later account, by contrast, was a philosophical pronouncement that bore the stamp of his 'meta-meta-physicking' conversations, as Coleridge dubbed them, about Kant's idealism and the redrawing of the boundaries between subject and object, mind and matter. The notion that the world is composed of 'impressions, ideas, pleasures and pains' was one that would have been familiar to most of the experimental subjects before the trials began, drawn as it was from David Hartley's theories of vibration and association, and it had underpinned the eloquent formulations of subjects such as Roget. But the conclusion that 'Nothing exists but thoughts' was one that had been reached by none of the experimental subjects, least of all Davy, prior to the composition of his researches for publication. Where Beddoes sought to convince the sceptics who suspected that the effects of nitrous oxide were all in the mind, Davy responded that the experiment, taken to its conclusion, rendered the distinction meaningless. Reality itself was constructed in the mind, from the information delivered by the senses: his culminating experiment had proved, as nothing ever had before, that an altered sensory and mental frame had the power to generate an entirely different universe.

7

New Worlds

During the first months of 1800, Britain was ravaged once more by the scarcity and unrest that many hoped had been left behind with the old century. Persistent and prolonged rain had ruined great swaths of the 1799 harvest, and by the spring of 1800 the wholesale price of wheat had doubled from the previous winter: across much of the country, the cost of bread exceeded the entire family budget. Around Bristol, peasant farmers turned to horse beans and turnips to keep themselves alive, eked out with potato peelings scavenged from those who had them to spare. In Cornwall, Giddy returned from assessing the county's grain stocks and ordering the brewers to spare their barley for food, only to find his own mill ransacked and his winter flour stolen.

As ever, the line between food riots and political unrest was hard to draw, not least because in this instance government policy was playing a primary role in creating the scarcity. The duke of Portland, as home secretary, had rejected proposals for quotas, price-fixing or other forms of state interference in the wheat supply, preferring to allow the market to direct supplies to the areas where the demand was greatest. But this policy had directed them to where they could be afforded, which was not the same thing; large sections of the country, such as the textile towns of the north, were boiling over in riots and mass popular unrest. During 1799 Pitt had introduced further measures to suppress dissent, including specific legislation to outlaw political societies and a Combination Act to prohibit the assembly of workers for negotiation or group action; their effect was to turn workers demanding bread into criminals and seditionaries. Mutinies ripped through a navy where rations had been cut to starvation levels, and waves of strikes and lockouts swept across the manufacturing towns, where crowds queuing for food were infiltrated by agitators whispering that the government was deliberately starving them, and their only recourse was to rise up against the king and join with the republican army of France.

With scarcity came disease. Typhus and other putrid fevers had long been endemic in the low-lying marshy areas around Bristol; as winter dragged on and food became scarce, they reached epidemic levels. In the densely packed slums and docks, whole families, streets and even districts were struck with fevers and rashes, chills and muscle pains. As the death count began to rise, Beddoes redirected the Pneumatic Institution's resources to emergency relief, publishing a handbill of instruction on hygienic practices, stressing the importance of washing after contact with sources of contagion, and keeping houses clean and well ventilated. As he tramped the docks and slums circulating copies among the city's poor, he observed that 'the number of cases was prodigious': at one address he saw '28 people lay down with fever', and at another 'eight were buried out of a single house'.[1] The experiments in pneumatic therapy, introduced only cautiously to begin with, now took a distant third place behind the victims of the epidemic and the conventional treatment of the syphilitic and the scrofulous. They were further eclipsed, too, by another novel therapy catching the public imagination: Edward Jenner had begun giving weekly smallpox vaccinations in Bristol as part of a programme spreading rapidly around the country, and even, thanks to the proselytising energy of the ship's surgeon Thomas Trotter, being adopted by the navy. Beddoes' scepticism regarding vaccination, in which he had until recently been among the majority, was now coming to seem like a conspicuous holdout: even Sir Joseph Banks and the Royal Society were now warmly supporting the extension of experimental trials.

Beddoes was also occupied with the procession of seasonal sufferers who had descended on him through the course of the winter. Gregory Watt, who had travelled south and west for the mild climate, stayed in Bristol for Beddoes' ministrations and the society of Davy. Robert Southey was suffering once more from interrupted heartbeats and palpitations, and moved from his rural cottage into the city to be near Beddoes who, with Davy, plied him with nitrous oxide and port. Tom Wedgwood, too, had returned to Beddoes' care for the winter. In August 1799 he and Jos had finally found the country home for which they had been searching: Gunville House, a country estate in the chalk uplands of Dorset, centred on a large mansion where Tom could convalesce in peace and Jos could join him in between overseeing the family business in Staffordshire. Tom had fallen in love with Gunville and the solitude of the rolling and sparsely populated Dorset hills, but by winter he was too ill to remain without a resident doctor, and had decamped to Rodney Place.

It seems likely that, over the winter of 1799–1800, Tom became an experimental subject for Beddoes' idea of cow-house therapy. 'I have tried living with cows in desperate cases of consumption with much promise of success in three out of six cases', he told Giddy in December 1799, presenting him with a plan for taking the project further. Beddoes envisaged 'a building for the reception of patients', a room warmed by stoves and adjoining a cattle-stall, so that the cows might poke their heads through a curtain into the patients' rooms and fill them with their breath.[2] 'For mere temperature living with cows is the most delicious thing imaginable', he enthused to Jos;[3] and, though he remained unsure of the chemical principle behind it, he was hopeful that 'the fumes would give a satisfactory stimulus to the surface of the lungs which might communicate itself to the whole system'. A patient could spend days, perhaps weeks, in a warm room breathing in the animal vapours that seemed to keep butchers so ruddy-cheeked and free from consumption.

Beddoes approached Giddy for a loan of 'from £400 to £500' to fit up a specialist cow-therapy ward, presumably on the ground floor of Dowry Square to adjoin a cattle-stall to be constructed in the yard behind the house. 'When I ask anything of a man he has as good a right to refuse as I have to ask', he reminded Giddy chivalrously, and it was a right of which Giddy seems to have availed himself on this occasion.[4] The case with which he was most familiar, that of his friend Lydia Baines, had not been among the promising ones: not only did the experiment 'produce no effect', but it profoundly unsettled the patient, who became 'very averse' to Beddoes' advice as a result. She had not shared his view that the presence of cows in the sickroom was wholesome, and had been horrified by the 'cowdung etc.' involved: the whole episode had convinced her that Beddoes was experimenting on her in his own interest rather than in hers.[5] But Miss Baines' unfavourable response had not deterred Beddoes; rather, it had suggested to him the refinement of a curtained stall that might allow the patient to receive the cow's breath without its other excretions.

Beddoes' cow-house therapy would become prominent in his posthumous mythology, usually in droll and parodic form, and more often than not in service of the view that his career had been studded with eccentric follies. This was a process that probably started with Maria Edgeworth, who passed on the story with the raised eyebrow with which she had witnessed the nitrous oxide experiments: 'one of his hobbies', she related of Beddoes, 'was to introduce cows into the invalids' bedrooms, that they might inhale the breath of the animals, a prescription which naturally gave umbrage to the Clifton lodging-house keepers'.[6] This snippet, with its mischievous image of Beddoes leading cows up and down the staircases of his patients' boarding houses, was picked up and further embroidered by others, notably Thomas de Quincey in his gossipy account of Coleridge's chequered medical history, and reaching as far forward as the literary essays of Lytton Strachey. But the image was passed on without the context that had led Beddoes to the practice. He was beginning to recognise that, for pneumatic therapies to be as effective as those received via the skin or stomach, they needed to be administered constantly over a long period, rather than in short and intermittent sessions. The growing success of vaccination, too, suggested to him that the vapours of cows might have benign effects in strengthening the system against infection. But more persuasive than anything else, perhaps, was the fact that Tom's condition was worsening. No therapy up to this point had produced anything beyond temporary relief, and it was only a matter of time before his exhausted frame succumbed to terminal collapse.

* * *

As winter turned to spring, Beddoes encouraged his patients to emerge from hibernation and to migrate to warmer climates. Southey's anxiety about his condition had abated but was being replaced by anxieties about his dependence on a regime of gases and alcohol; he took Beddoes' advice and returned to Portugal, taking his protesting wife, Edith, with him, and leaving Davy with a sheaf of poems and other contributions for a further *Annual Anthology* that would never materialise. Tom Wedgwood, casting around ever more desperately for a cure, took the drastic step of leaving the care of his family to sail for the West Indies. Jos, with a wife and young children, was unable to accompany him; he had never been so profoundly separated from Tom, to whom he wrote that 'I have not been able yet to think of you with dry eyes'. But he was well aware that even such an unhappy and risky avenue needed to be explored, since 'your admirable qualifications are rendered ineffectual for your happiness by your miserable health'.[7]

Tom's first impressions of the Caribbean exceeded his wildest expectations. The prospects of 'little valleys at the feet of mountains piled on each other in a noble succession, every tree new to the eye, many loaded with brilliant flowers and fruit', filled him with wonder and vigour, and with the hope that he had 'found a paradise' in which he might finally be restored to health. He reported that he was beginning to 'gain strength very rapidly' and, with characteristic stoicism, wrote to his family that 'if I had no indigestion and headache I should be in heaven'. But he had learned to be vigilant against the enervating cycle of euphoria and despondency, and added the caution that 'I dare not indulge in the luxury of those feelings which begin to introduce a disordered agitation'.[8] It was a pattern that was becoming tragically familiar; and indeed, just as with his nitrous oxide trials, his body would allow him only a short interlude of relief before reminding him who was master.

This was a pattern that was also beginning to emerge more generally from the Pneumatic Institute's experiments, where attempts to capitalise on the early promise of the first palsy treatments were proving problematic. The pneumatic trials had begun sporadically and tentatively while the Institution established itself among its intended clientele, and once nitrous oxide had revealed its wildly unexpected effects on consciousness, the trials had shifted their focus towards the willing and able-bodied subjects who jostled to experience its novelties. But now that the typhus epidemic had struck, it had proved harder than ever to extend them. The therapeutic trials had generated a handful of tantalising case histories but also a mounting and largely unrecorded procession of failures, leaving the evidence for the medical power of the airs inconclusive at best. Although it could still be argued that they had not had the full and fair trial that the Institution had been established to guarantee, patients had already begun to vote with their feet. They were happy to queue for vaccination, for which trials and conclusive results were multiplying across the country, but not to volunteer for a pneumatic cure of which patients and doctors alike were becoming increasingly sceptical.

There were various sources for this scepticism. Joseph Cottle recalled that 'an idea had become prevalent amongst the crowds of afflicted, that they were merely made the subjects of experiment, which thinned the ranks of the old applicants, and intimidated new'.[9] It was a suspicion to which Beddoes had been alive from the beginning, though he also suspected that it was being maliciously orchestrated, part of a pattern observable 'whenever attempts have been made to deliver mankind from the plague of sickness'.[10] He was certainly receiving little support from the Bristol Royal Infirmary – perhaps unsurprisingly, given that he had accused it in print of operating an 'illiberal and unjust' cabal whose policy of electing physicians for life excluded the majority of promising junior and trainee doctors.[11] The Institution was also dogged by its Jacobin associations. When, for example, a shipment of frogs destined for Davy's laboratory escaped on Bristol's quayside, the rumour swept through town that they were intended as food for French revolutionaries hidden in the cellars beneath Dowry Square. Beddoes had also, apparently, become a target for the black man whose hand he had bleached several years previously in

Oxford, and who was now 'sufficiently artful to station himself in the neighbourhood of the Institution, and to appeal to the compassion of all he met, by telling
them that he had been decoyed thither, and made, without his knowledge, the
subject of a cruel and unheard of experiment'.[12] If Beddoes believed that there
were those conspiring to ruin his reputation, he was not without plausible
grounds.

For Cottle, whose later narrative of Coleridge and Southey's Bristol years
presented the Institution as a frothy episode of comic relief, the reason for its
dwindling supply of volunteers was simply that the pneumatic cures did not
work: 'it is too great a tax on human patience, when cures are always promised,
but never come'.[13] But the facts were more complicated. Patients' expectations
had been raised by the early cures, often to implausible levels. One early para-
lytic patient, Coleridge recalled, was prepared for experiment by Davy placing a
thermometer in his mouth; he immediately pronounced that 'the talisman was
in full operation', and that 'he already experienced the benign effects of its influence throughout his whole body'.[14] This anecdote, furnished to Davy's biographer a generation later, may have been embellished for comic effect, but it
illustrates the difficulty of assessing the results of an experimental therapy
conducted only as patients and circumstances allowed.

If the therapeutic evidence had proved inconclusive, so too had the evidence
for the mechanisms through which the gases operated on the body. Nitrous
oxide, in particular, seemed to inhabit a dimension that ignored the Brunonian
dichotomy between stimulant and sedative. On the one hand, it excited the
nervous system, raising pulse and respiration as it took hold of the body, stimulating powerful muscular spasms and, in some therapeutic cases, imparting an
energy to the exhausted or palsied patient that no other treatment could match.
On the other, in many cases it seemed to dull the nerves, blunting sensations
of pain, propelling subjects into a swooning trance and even, on occasion, into
prolonged states of lassitude where their nervous power was sapped for hours, if
not days. Davy's heroic attempts to resolve this question had revealed still more
extraordinary effects, leading him to a world where nothing existed but thoughts,
but had brought him no nearer to an answer. It was a conundrum that devolved
into further questions about how precisely respiration worked, which gases were
involved, and how they were absorbed into the blood; and beyond these, into the
deeper mysteries of life itself and the means by which it was animated.

* * *

But if the therapeutic effects of the gas had fallen short of expectations, word of
the philosophical experiments of Beddoes' circle had carried further than he

had anticipated, and in May 1800 they made their literary debut in the form of an extended satire in the government-sponsored journal *The Anti-Jacobin Review*. Established by George Canning during the militia drive of 1798, this publication had served an important role in rallying demoralised public opinion in the face of mutinies and fears of invasion, and had subsequently become the establishment's journal of record: Pitt and other members of his administration contributed anonymously, and the caricaturist James Gillray had been lured to turn his prodigious and obscene talents to patriotic propaganda. It was an influential compendium of news and reviews, generously peppered with scabrous satire, of which 'The Pneumatic Revellers: An Eclogue' was a typical example. The scene it depicted, grotesquely but not entirely implausibly, was of a circle of self-important philosophers holding a gas-fuelled symposium of gibberish in exchanges of doggerel verse.

As in *The Golden Age*, Erasmus Darwin was taken to be Beddoes' tutor in atheism and novelty, and the skit drew its theme from his famous prediction in *The Botanic Garden* that mankind might one day domesticate the power of wind and waves, and learn 'to ride the whirlwind, and direct the storm'.[15] Now Beddoes' new gas, announced 'The Pneumatic Revellers', 'has shown how far a philosopher may be carried by the force of a flaming imagination'.[16] Beddoes, as ever a gift to satire, allows the anonymous author (perhaps the Cornish clergyman and poet Richard Polwhele) to open with snippets of authentic reportage plucked verbatim from his *Observations*, from the unguarded eccentricity of announcing himself 'bathed all over with a bucket of good humour' when he made his first self-experiments, to his grandiose prediction of a revolution in the human condition by which we might control the ultimate sources of pleasure and pain, and perhaps even dispense with the need for sleep. 'Such stores of health and pleasure, has Dr. B in reserve for his fellow-creatures!' the author marvels. 'So wild is my wonder, so intense my gratitude, in the contemplation of a philosopher to whom Newton is an ape . . . that I can add no more!'

But there is plenty more, as the sketch turns into a round of drinking songs, with each philosopher attempting to outdo the last in his praise of the gas. The poet George Dyer, a Unitarian friend of Coleridge who had joined the experiments and was already a familiar *Anti-Jacobin* target thanks to his tracts on the plight of the London poor, leads the toasts with

Beddoes! Thy living beverage whilst I quaff
I laugh! – ha! ha! – yet know not why I laugh!

Robert Southey, equally familiar to the journal's readers as a pretentious republican hothead, caps this with a soliloquy:

> I am rapt beyond myself! I feel
> At my extremities delicious thrillings!
> My every sense is exquisitely keen!
> My taste is so refined. . . .

At this point Beddoes himself, whose girth was well enough known to need no
explanation, takes up the narrative:

> If at dinner indeed I indulge in much merriment
> And dispatch a sirloin, 'tis by way of experiment.

He concludes his disquisition on the experiments conducted 'amid chemic
inanity' by congratulating himself on the opportunity he has given the distinguished company

> To rejoice in an air from corruption so free
> As the gas, my good sirs, just emitted by me.

The target is broad enough, and hit repeatedly and with relish. There are
numerous plays on the subjects 'giving themselves airs', the business 'getting
wind' and the company being 'blown up' with the pomposity that the satire
gleefully pricks. There are abundant reminders, too, of pneumatic chemistry's
Jacobin hoof ('By Satan, or Priestley prepared, 'tis all one'), and comparisons to
the wildest projections of empirics, quacks and alchemists ('a discovery to
shame the Philosopher's Stone'). But beneath the predictable jibes the caricature was almost one that Beddoes and his circle might have produced
themselves in a moment of levity: the incongruous mingling of profundity
and absurdity, and the bathos of the descent from high-flown metaphysics
to antic hilarity, was a trope with which they had made much play themselves.
By the same token, behind the crude and ribald house style the trademark
Anti-Jacobin venom is softened to a genial absurdity, even perhaps a sense
that the author might have liked to be able to join in with the tomfoolery.
Preposterous the Institution might be, but it was hard to construe its
activities as sinister. The journal had spent two years warning of the deadly seri-
ousness of a Jacobin threat gathering in the shadows; but if this was the extent

of the revolutionary plotting of 1799, the threat had plainly dissolved into hot air.

*   *   *

But the definitive record of the experiments was yet to emerge and, when it did, it would prove considerably more resistant to mockery than Beddoes' *Observations*. From his climactic self-experiment on Boxing Day 1799 until Easter 1800, Davy had been hard at work finishing up his experiments with nitrous oxide and other gases; from then until midsummer he wrote them up at a furious pace. By the end of the summer they had emerged from the press, under the imprint of Joseph Cottle in Bristol and Joseph Johnson in London, as an exhaustive account, at nearly six hundred pages twice the length of anything Beddoes had ever written, under the title *Researches Chemical and Philosophical, Chiefly concerning Nitrous Oxide, or Dephlogisticated Nitrous Air, and its Respiration*. Where Beddoes' *Observations*, like his entire oeuvre, had been allusive and opinionated, full of vigorously ridden hobbyhorses and generous in Shandean diversions, Davy's *Researches* was tightly methodical and exhaustive in its detail. It lacked Beddoes' lightness of touch – long stretches seemed designed to do little more than beat the reader into submission – but it dealt a sizeable blow to any who were hoping to dismiss the experiments as a grandiose folly. As framed by Beddoes, the effects of nitrous oxide were a curiosity, a bizarre and unlooked-for outcome of medical trials that had collapsed into unanswerable questions and hurriedly configured follow-up projects; as reframed by Davy, they were the climax of a research programme that had carefully recorded every step of its progress, and in doing so had laid firm foundations for interpreting even this most unexpected outcome.

Davy's book comprised, as its title implied, a series of parallel researches that built one upon another to present a narrative that ran seamlessly from the fundamental building blocks of chemistry to the visionary revelation of a universe composed entirely of thoughts and ideas. The first set of researches analysed the combinations of nitrogen and oxygen – including nitrous oxide, nitric oxide and nitric acid – in grinding detail, recording their specific gravities, the vapours they gave off when heated, the combinations they formed with water, and their rates of absorption by charcoal, iron and other metals. These demonstrated more clearly than ever before the variety of forms that could be assumed by two gases that had been thought until recently only to exist as part of the neutral backdrop of air. Not only did each gas have hitherto unsuspected and complex properties, but the variety of their combinations spoke to the unimagined complexity of the imponderable forces, fluxes or bonds that must be capable of existing between them.

The second researches moved from the general principles of these gases and their combinations to nitrous oxide in particular, exploring the entire range of its chemical behaviour: its combinations with saline solutions, its reactions with soda and potash and with other gases, and its decomposition by those substances such as sulphur and hydrocarbonate that had the power to dismantle its bonds and reduce it to its component parts, or to configure them into new combinations. The third researches brought nitrous oxide to bear on the study of respiration, recording the changes in metabolism that it induced in animals, fish and insects, its absorption and dissolution in blood and the effect on different organs of immersion in it. Finally, the fourth researches, 'Relating to the Effects produced by the Respiration of Nitrous Oxide on Different Individuals', introduced the human experiments, with the effects on body and mind recorded in their own words by over thirty of their subjects. As Davy was quick to concede, these descriptions were imperfect, perhaps even incomprehensible, but the problem was not in the method or coherence of the researches; rather, it was inherent in the effects of the gas and the limitations of the 'language of feeling' available to describe it. 'We are incapable of recollecting pleasures and pains of sense', he observed, and 'it is impossible to reason concerning them, except by means of terms which have been associated with them at the moment of their existence'.[17] The reports were undeniably strange, but only at a trivial level should they prompt ridicule.

Davy's reframing of the researches brought the nitrous oxide experiments into the purview of scientific investigation, but in doing so it had the effect of marginalising the questions they had been designed to answer: the effectiveness of pneumatic medicine and, particularly, the validity of the Brunonian system. The lines of enquiry that he followed were essentially chemical rather than medical, inductive rather than speculative; he had begun not with Brown's theories but with his exhaustive trials of the gases themselves, and they had led him away from the broad and simplistic categories of excitation and weakness. When he finally turned to Brown's theories in the conclusion of the book, his assessment was brief and, in its implications for Beddoes' project, devastating. Brown's theory demanded that a gas such as nitrous oxide have a fixed position on the sliding scale from stimulant to sedative, and that by introducing it into respiration the doctor should be able to shuttle the patient up and down this scale in an orderly manner; but the experiments had shown that this 'common theory of excitability is most probably founded on a false generalisation'.[18]

If it was indeed a false generalisation, it collapsed the entire edifice of pneumatic medicine as it was currently conceived. On a physiological level nitrous oxide's effects radiate out from the lungs through the whole organism, whereas most diseases are characterised by the malfunctioning of a particular organ: if factitious airs should prove to work in therapy, they must do so by a principle as yet undiscovered. 'Pneumatic chemistry in its application to medicine', Davy concluded in strikingly harsh terms, 'is an art yet in its infancy, weak, almost useless'. He supported Beddoes' testimony that nitrous oxide had produced remarkable early results in palsy, speculating that 'as by its operation the tone of irritable fibre is increased . . . it is not unreasonable to expect advantages from it in cases of simple muscular debility'.[19] But the infant art of pneumatics still needed 'to be nourished by facts, strengthened by exercise, and cautiously directed in the application of her powers by rational scepticism'.[20] It was the hallmark of true experiment that no one should be able to predict its outcome, and that it should have the power to demolish the theory it had set out to prove; in this case, it might be said, the operation was a success, though the patient had died.

The nitrous oxide researches had demanded sacrifices of theory from Davy as well as Beddoes. Davy had begun them wedded to his notion of 'phosoxygen', derived from his conviction that heat, light and combustion were all manifestations of the same principle, which was in turn the principle underlying all life; and he had initially theorised that the key to nitrous oxide's effects was that it 'contains more light in proportion to its oxygen' than any other gas.[21] He had committed his theory to print, not only in the *Essays on Heat and Light* published by Beddoes but also in his communications to *Nicholson's Journal*, the bulletin of record for British experimental chemistry to which, with his eye as ever on claims of priority, he had been contributing letters announcing his work-in-progress. But in February 1800 he had dashed off a retraction to *Nicholson's*, confessing that 'I beg to be considered as a sceptic with regard to my own particular theory of the combinations of light'.[22] It was an embarrassing climbdown; and there were more sacrifices to come, notably his conviction that air was not a mixture but a compound. Yet it was Beddoes who was hit hardest by the conclusion of the *Researches* and, as its success mounted, its effect was not only to establish Davy as a groundbreaking chemist but also to crystallise the perception of Beddoes as a man whose project had failed the test of experiment on which he himself had insisted so loudly and so long.

Davy had not set out deliberately to sabotage Beddoes' project, nor is there any suggestion that Beddoes felt he had. Both were wedded to the experimental method, and thus entirely prepared, at least in theory, to submit their most cherished beliefs to it. The previous year Davy had described Beddoes as 'the most truly liberal, candid and philosophic physician of the age', firmly wedded to experiment and able to give up his theories 'whenever they appear contradictory to facts'.[23] If Brunonianism were false and Beddoes' Baconian principles were genuine, he would wish to know it. But Davy was well aware that his conclusions were a betrayal, if not of Beddoes, then of the project that had allowed him to launch his career. His notebooks contain a draft of a gushing dedication to Beddoes – 'without you the researches detailed in this volume would probably never have been made'[24] – but the printed version finds a revised and far cooler acknowledgement, stating simply that 'I have been aided by his conversation and advice', and that the researches 'were executed in an institution which owes its existence to his benevolent and philosophic exertions'.[25] The researches had begun with Beddoes and his theories the driving force, and Davy his assistant in service of their common goal; but by the time they had ended, Davy knew that his work had transcended them.

Davy's *Researches* marks the beginning of the eclipse of Beddoes' dream of pneumatic medicine. Although he never entirely recanted his theories, and continued to maintain that they deserved further trial, neither did he mount any public challenge to Davy's conclusion: he submitted a brief appendix to the *Researches* in which he acknowledged that 'to suppose that the expenditure of a quality or substance or spirit, and its renewal or accumulation are the general principles of animal phenomena, seems to me a grievous and baneful error'. Brunonianism, and indeed all theories of life, would remain at best provisional until the essential principles of physiology were better understood. He had always been diffident in defending Brown, but from this point his appetite for doing so diminished markedly, and his thoughts on the physiology of life tended increasingly to run down other channels. But behind this silence lay the impact of a cruel blow, made crueller still by the astonishing early promise of the experiments. The project, for Beddoes, had turned out to mirror curiously the effects of nitrous oxide itself: a racing of the pulse as the excitation began to take hold, building to a thrilling sense of imminent revelation as it reached its peak, before the air-bag deflated and the vision receded as quickly as it had arrived, leaving only tantalising shards impossible to assemble with any conviction, and a chorus of raspberries from the assembled spectators.

* * *

Thanks to a curious artefact of hindsight, however, the main reason for which Davy's *Researches* is remembered today has nothing to do with Beddoes' agenda, and indeed little to do with Davy's. One sentence on page 556, tucked away at the end of a series of speculations on possible medical uses of nitrous oxide, has attracted more subsequent attention than the rest of the book put together. After considering its usefulness for resuscitating unconscious or debilitated subjects, or in extended doses for giving vigour to patients with weak constitutions, Davy suggests that 'As nitrous oxide in its extensive operation appears capable of destroying physical pain, it may probably be used with great advantage during surgical operations in which no great effusion of blood takes place'.[26]

As one among many speculative and unevidenced proposals for medical uses of the gas, and with far more spectacular effects to claim the attention of the public, this comment attracted no contemporary interest whatsoever. But fifty years later, once nitrous oxide had been demonstrated to be effective in precisely this context – and the American writer and physician Oliver Wendell Holmes had coined the term 'anaesthesia' to describe a state where pain was not merely relieved (analgesia) but consciousness suspended entirely – Davy's suggestion posed a problem. Why, if this miracle of modern medicine had been proposed half a century previously, had it taken the medical profession so long to implement it? The answer, in the form in which it was resolved in the second half of the nineteenth century, has persisted largely unexamined into the medical textbooks of the present day: Davy and Beddoes, preoccupied with agendas that are now known to have been dead ends, 'missed' the anaesthetic properties of nitrous oxide, or failed to recognise their implications. As a result, the work of the Pneumatic Institution was forgotten, and the world was obliged to wait for the defining application of pneumatic medicine to be rediscovered.

The posthumous reputation of the Pneumatic Institution, such as it is, is almost entirely wrapped up with this debate, which has had a curious distorting effect on the later understanding of the project. On the one hand, it identifies Beddoes and Davy as heroes of science by pointing to the eventual vindication of their researches: despite the almost universal disbelief of his contemporaries, Beddoes would prove entirely correct in his predictions that the factitious airs would transform medicine, and in particular the control of pain. But what the subsequent discovery of anaesthesia gives with one hand it takes away with the other: by highlighting the gap of fifty years between this work and the heroic moment of medical triumph, it implicitly blames Beddoes, or Davy, for the failure to seize what would turn out to be their chance to transform the art of medicine. Furthermore, if the Institution's claim to fame rests solely on the idea that pneumatic medicine might be used to counter the pain of surgery, credit should rightfully be given neither to Davy nor to Beddoes, but to Davies Giddy, who had committed this idea to writing five years before their researches had even begun.

All these retrospective allocations of vindication or blame, however, rest upon the possibility of an alternative scenario: that Beddoes and Davy might have rushed into print to propose nitrous oxide as an adjunct to surgery, and then devoted themselves to campaigning for it until the medical profession was swayed. But there are many reasons why this sequence of events was impossible. First, the effect of the gas on most of the Institution's subjects was precisely the opposite of anaesthesia. No surgeon, then or now, would choose to operate on a patient who was stamping, shouting and laughing uncontrollably, running around the operating theatre or seized with involuntary muscular spasms. The dose used in surgery is a good deal higher than the contents of a green silk bag, and administered by an apparatus that can deliver it continuously; but the contrast is also a reminder that subjects surrounded by excited onlookers and expecting a poetic epiphany will respond very differently from patients on a trolley attended by hushed and masked professionals.

Even if Beddoes and Davy had somehow decided to throw all their efforts behind encouraging the use of gas in surgery, it is hard to imagine how they could have succeeded. The surgeon among their volunteers, Stephen Hammick, who grabbed the air-bag violently under the influence of 'the strongest stimulant I ever felt', would doubtless have gone to great lengths to keep similarly intoxicated patients as far away from his surgery as possible. Beddoes had been raining chemical treatments on the medical profession for years, and tirelessly announcing the benefits of pneumatic therapies, with only a carefully cultivated handful of physicians prepared to support his cause. If the profession was unwilling to experiment with gases even for the symptomatic relief of invalids with incurable conditions such as consumption, it is hard to imagine experiments being sanctioned in the surgery, where furnaces, chemical reactions, naked flames and silk bags all presented life-threatening obstacles to critical procedures.

There were broader cultural reasons, too, why the elimination of pain in surgery should have seemed fanciful in 1800, but the crowning glory of the physician's art in 1850. The intervening years were ones in which the notion of pain was itself transformed, from a regrettable but necessary fact of life to an affront to human dignity from which patients should be spared wherever possible. Davy himself was a case in point: as a young man, adventurous, lucky and accident-prone in equal measure, he would often insist that 'a firm mind might endure in silence any degree of pain, showing the supremacy of "mind over matter"'.[27] Although he mentions nitrous oxide's ability to dull pain at several points in the *Researches*, it is unsurprising, especially given the far more dramatic effects he was investigating, that he did not pursue or prioritise it.

Davy's attitude to pain was more typical of its time than that of Coleridge, who was in the vanguard of the revolution in sensibility from which the aspiration to pain-free surgery would eventually emerge. It was a subject about which he had a perhaps morbid curiosity, in which he felt he was unusual. 'I want to read something by somebody expressly on pain', he wrote to Davy at the end of 1800; 'it is a subject that exceedingly interests me'.[28] The increased resort to opium in the 1790s, both by doctors such as Brown and Darwin and patients such as Coleridge, would come to be seen as an early indication of a new sensibility towards pain; but the ambition that Beddoes expressed for medicine to 'come to rule over the causes of pain' would be more controversial and sharply criticised than almost any other aspect of his project. It would not be until the 1830s that doctors and surgeons would begin in earnest to challenge the religious position that pain was 'the voice of nature', a necessary condition of life – and the medical view that pain was a stimulus that kept traumatised patients alive – and to regard the elimination of pointless suffering as one of medicine's duties. Anaesthesia, when it emerged in the 1840s, was as much a response to surgeons' needs as to patients': technical advances had led to more sophisticated operations, and the ability of the patient to endure them had become a limiting factor that needed to be addressed.

Yet even at this late stage, much of the medical profession remained, by modern standards, remarkably incurious about relieving pain. Many surgeons rejected the very idea of anaesthesia, insisting that the crucial elements in an operation were the surgeon's skill and the patient's bravery, and that elaborate chemical interventions were more trouble than they were worth. After it was pioneered in America in 1846, the celebrated British surgeon Robert Liston would characterise anaesthesia as a 'Yankee dodge' that undermined surgery's time-honoured traditions; although Liston recognised its value, others continued to maintain that 'pain is a wise provision of nature, and patients ought to suffer while the surgeon is operating'.[29] The Royal College of Surgeons would not approve nitrous oxide anaesthesia until 1868.

Standard histories of anaesthesia, as of medicine in general, have a tendency to assemble themselves from a procession of 'eureka' moments that record the first appearance of modern technologies, and in doing so encourage the assumption that their modern applications were promptly recognised and adopted by the profession. But such assumptions are frequently problematic. When anaesthesia finally entered surgical practice via dentistry, for example, it was not nitrous oxide that won the race to convince surgeons of its utility but ether, the volatile inhalant with which Beddoes had been treating Southey the previous winter. This was a treatment that Beddoes had learned from Darwin, but ether's discovery dates back to medieval Arab alchemists, and its use as an analgesic to Paracelsus, who recorded in 1525 that inhaling 'stupefying vitriol salts' – most likely ether – made chickens 'fall asleep, but wake up again after some time without any bad effect', and 'extinguishes pain' for the duration.[30] The discovery of ether anaesthesia was attended by a famously bitter contest for priority, with claim and counter-claim dragged through the courts;[31] but there was, in theory at least, no reason why surgeons could not have been using it during their operations for centuries. Nitrous oxide's discovery, by contrast, depended on the modern synthesis and administration of factitious airs; as a result, Davy and Beddoes' claim to priority was too close in time and in scientific intent for the surgeons of the 1850s to ignore entirely, and thus needed to be either acknowledged or disposed of.

For all these reasons, the attempt to connect the Pneumatic Institution's work to the subsequent discovery of nitrous oxide anaesthesia, and to declare it a success or a failure in consequence, is one that offers little insight into either story. Davy's suggestion can perhaps be best understood simply as premature, as is any discovery, in Gunther Stent's classic formulation, 'if its implications cannot be connected by a series of simple logical steps to canonical, or generally accepted, knowledge'.[32] The nitrous oxide researches, in this sense, simply emerged at a time when the benefits of their defining application had not yet been grasped. Just as Gregor Mendel's discovery of the hereditary unit in 1865 languished unappreciated until the idea of genetics had taken hold – or, indeed, John Mayow's achievement in isolating oxygen could not be fully grasped until Lavoisier had named it as an element a century later – the Institution's nitrous oxide researches would remain for decades simply an anomaly, a wonder or a folly that could be marvelled at or ridiculed, but could not be evaluated or applied until the world was ready for them.

*   *   *

Beddoes, in the meantime, had more to worry about than the opinion of posterity. In August 1800, lest perhaps he should feel that he had got off lightly with the gentle ridicule of 'The Pneumatic Revellers', The Anti-Jacobin made space for a second attack on him, this time a lengthy review of his Observations from the previous autumn. It began with a dry account of the synthesis of nitrous oxide, full enough for any chemically knowledgeable readers to reproduce the gas themselves; having thus discharged its obligations to fairness, it proceeded to a savage attack on the arrogance of Beddoes' claims. Like 'The Pneumatic Revellers', it seized on his prediction that humanity might eventually come to rule over the causes of pain and pleasure, and announced its eager anticipation of this 'complete revolution' where 'we shall be made immortal in the twinkling of an eye, or rather we shall be made over again; for we are to receive new bodies and new minds too; frogs are to be converted into oxen; and oxen, no doubt, into men'.[33] Indeed, if the properties of the gas were as miraculous as he claimed, it urged Beddoes simply to convert the entire atmosphere to nitrous oxide: 'this would make us all angels in a trice; not to mention the inexpressible pleasure of being drunk our lives long'. Given such prospects, 'to cure the palsy is a mere trifle, scarce worth mentioning when compared with the immense magnitude of our author's projects'.

It was these contrasts – between the grandiose claims of the Institution and the tragic reality of serious illness, and between what Beddoes asks us to accept and what he has actually proved – that led the reviewer to his serious charge. Palsy is an agonising and chronic condition, and it is no light matter to claim to have found a cure for it; but Beddoes has done so on the basis of two experiments, raising the hopes of thousands of sufferers while, in truth, having nothing to offer them beyond more projection and sales pitch. Simultaneously, he has backtracked with breathtaking insouciance from the claims of a cure for consumption that he has been touting incessantly for the past five years. In fact, 'disappointments seem rather to have increased than diminished his confidence of ultimate success': each failure is accompanied by more grandiose expectations and more claims of imminent revolution. A responsible physician would take into account 'the wonderful effects which all new remedies produce, while assisted by their novelty and the enthusiasm of the discoverers and their patients', and would revise their claims downwards rather than upwards as a result.

But this is not how Beddoes proceeds, the reviewer continues, and the reason is a fundamental flaw in his approach to medicine. He is hypnotised by the prospect of the technical breakthrough, the chemical fix, the miracle cure; but this prospect is an illusion. It stems from a belief that the secrets of life are on the verge of being unlocked; but the truth is that 'the phenomena of life' are infinitely more complex than 'the heated brains of modern theorists lead them to suppose'. They cannot be reached via 'the acids, the alkalis and the fermentations of the chemists', nor reduced to the single principle of excitation peddled by John Brown. Medicine is not the province of 'men of genius, as Dr. B understands the term', who can solve all its mysteries with a flash of inspiration; rather, it advances by small increments down 'the rough road of patient investigations'. 'New and gaudy paths of hypothesis and conjecture' have always been with us; but they have only ever led to inflated claims built on ever more rickety foundations. 'Let us hear no more then of new paths', the reviewer concludes; it is not merely the pneumatic project that is fatally flawed, but the entire spirit of Beddoes' investigation. 'Dr. B's men of genius have been at work, in all ages, with their pastes and painted glass of no value', and the physician is no better off today than if they had never existed.

It was a review that at least did Beddoes the courtesy of spelling out the objections of his shadowy detractors; and some of them must have hit their mark. He was not insensitive to the charge of raising patients' expectations, nor that of the dubious ethics of doing so to advance his own schemes, and thereby appearing in the same light as the quacks he despised. These were charges, too, that the nitrous oxide experiments had left him poorly positioned to answer. As much as the gas had been an extraordinary discovery, it had also become a maddening curse. Its effects, clearly fundamental in their implications, had lured him into making his triumphant announcement of the pneumatic revolution, then left him clutching an empty air-bag. To those who had experienced it with him, its powers were beyond doubt; to those who had not, it had generated claims simply impossible to believe, and left him exposed as never before to the scorn and ridicule of his detractors. It had become the tragicomic emblem, and perhaps the epitaph, of his life's work.

* * *

But although Beddoes seemed to have staked and lost his all on nitrous oxide, the Pneumatic Institution was far from finished. His commitment to untrammelled experiment had, as Priestley had warned, led him down an alley that had teemed with promise but proved unprofitable; but the same commitment had created the conditions under which further unsuspected and unplanned discoveries might emerge. They would not be long in coming. The summer of 1800 in Dowry Square would be as extraordinary as that of 1799, and it would deliver the revolution in chemistry that the gas had promised in medicine.

The summer began with more patients in residence. Davies Giddy, after a winter of riots in Penzance and Redruth that had eventually forced him to call out the county militia from Truro, was exhausted and ill; fearing that he was suffering the early stages of consumption, he came to Beddoes for treatment. He moved in to Rodney Place, much to the delight of Anna, who had never met her husband's oldest and closest friend. Given his weak state, Beddoes decided against treating Giddy with nitrous oxide, preferring to use digitalis, which he now considered a more useful specific against consumption. Giddy's treatment lasted two months, after which Beddoes dispatched him back to Tredrea with a clean bill of health, although uncertain of whether he had been suffering from consumption in the first place. He had begun to suspect a nervous preoccupation with illness in his old friend, even a degree of hypochondria that had grown, perhaps, from his discomfort with the responsibilities that his position in life had brought him. Yet it might equally have been that the disease was real but had been treated successfully in its early stages: successful preventive medicine, after all, made it hard to tell the difference.

In June, Tom Wedgwood returned from the West Indies, to the relief and delight of his family. They had all missed him deeply, and feared for the worst from the exertions of travel and the profusion of tropical diseases rampant in the 'torrid zone'. Tom had found the tropics gorgeous and endlessly fascinating, and had experienced moments when the heat had brought him back to vigorous health; but these had become fewer and further between, and he had found himself increasingly ground down by exhaustion and homesickness. 'My birds are singing on all sides of me', he wrote in his last melancholy letter to Jos from Barbados, 'oranges by thousands close to the house – a supper of land-crabs *in prospetto* – and yet I crave for . . . dear, dear Gunville'.[34] He returned to summer on his beloved Dorset estate, but he did so without a cure and, perhaps, having given up hope of ever finding one. 'Henceforth', he confided to Jos, 'I will never entertain, or at least communicate to others, these sanguine expectations of returning health'.[35]

Tom's return also brought a welcome addition to the Pneumatic Institution in the form of John King, a young surgeon whom Beddoes had selected as his companion in the West Indies. King was Swiss, and had changed his name from Johann Koenig when he arrived in London in 1791 to serve as an apprentice in the surgeon trade. He was a man of exceptional talents – besides his medical skills, he was very learned in mathematics and physics, and a fine copperplate engraver – but his abilities had thus far hindered rather than helped his career, combined as they were with a forthright atheism, a cause for which he had courageously spoken in public at the London Forum. Both his talents and his opinions were, however, welcome in Dowry Square, where he moved in to the attic bedroom next to Davy's and took on the role of resident surgeon. He soon proved not only a relaxed and confident physician to the Bristol poor, but also a far more methodical keeper of medical records than either Beddoes or Davy: from this point on, the Institution's books record not merely the promising or curious cases, but notes on every patient who received treatment. Finally, at the age of thirty-three, King had found a welcoming home where all his talents could be exercised. He cheerfully dissected frogs with Davy, engraved tropical scenes for Southey's epic poem *Madoc*, and within a year literally became a part of the family by marrying Anna's younger sister, Emmeline.

King's arrival gave Davy the opportunity to reduce his surgery and patient care duties, enabling him to devote himself more fully to his chemical experiments, and to explore more of the curious avenues that the nitrous oxide work had suggested. In his researches with the caustic and toxic 'nitrous gas', nitric oxide, Davy had noticed that it reacted with wet metal – zinc, copper or tin – to convert itself into nitrous oxide. This was a reaction that seemed to be a function not of heat, but rather of some kind of electrical process: as he had put it at the time, 'the action of the galvanic fluid'.[36] By the summer of 1800 a new technique had opened up for exploring this curious phenomenon further. In April, Sir Joseph Banks had received a letter from Alessandro Volta who had, since Galvani's discoveries of response in frog muscles that had electrified Beddoes in 1794, established that the source of the electricity was not (as Galvani believed) in the frogs' legs themselves but in the metallic contacts with which they formed a circuit. Combining layers of metal in a salt solution, Volta had found that he could not only build electrical charges but, for the first time, maintain a steady current. He had, he announced, now assembled an 'artificial electric organ', a mechanical counterpart to creatures such as the electric eel or torpedo, which bridged the crucial gap between chemical substances and the mechanisms of life.

This was a discovery that acted, as Davy put it, as an alarm bell across the scientific community. The knowledge of electricity in its static form was ancient: the name itself derived from the Greek for amber, a substance that, when stroked, would attract materials such as straw. The signature breakthrough of the previous century had been the Leyden jar, developed in 1745, which allowed the invisible electric 'fluid', as it was conceived, to be collected in a receptacle and discharged by bridging a spark gap. In 1749 Benjamin Franklin had shown that lightning was generated by the same principle of electrical discharge. But the Leyden jar delivered only a single jolt, which immediately emptied it; Volta's 'pile', as it was becoming known, could 'act incessantly, and without intermission'.[37] This was a development that had long been pursued, and one that many anticipated might resolve the great questions of the chemical properties underlying light and heat. Joseph Priestley, who had experimented extensively with the Leyden jar, had predicted in 1767 that this was a field where 'the bounds of natural science may possibly be extended, beyond what we can now form an idea of'. In a passage that had tantalised experimental science for a generation, he had prophesied that 'new worlds may open to our view' which might even eclipse 'the glory of the great Sir Isaac Newton'.[38]

This was the destiny of which Davy had dreamed ever since he had entered Newton's name alongside his own in his first notebook in Penzance; but he was not alone in his ambition. A translation of Volta's letter had appeared in the Royal Society's *Proceedings*, and William Nicholson had published his own preliminary findings in the June edition of his *Journal*; by July half a dozen researchers had published papers or letters on their experiments. Nicholson, in collaboration with his partner Anthony Carlisle, had shown that water, when exposed to the 'voltaic pile', separated into its components, oxygen and hydrogen, one being liberated at each pole. Electricity, it seemed, when controlled and flowing in a regular current, was, like heat or light, a force that could break substances down into their constituent parts, and reveal the elemental building blocks of which they were composed. It was still not clear precisely how the pile worked; but it was certain that the answer to this question would expose the workings of matter on a more fundamental level than had thus far been possible.

Whatever differences Beddoes and Davy may have had over the nitrous oxide researches, they now plunged together into a furious programme of voltaic experiments, racing throughout the summer against the pack of experimenters who had snatched a head start. Their combined skills made them a formidable team. In the Institution's laboratory they set to constructing a massive pile of a hundred alternating plates of zinc and silver; Beddoes, meanwhile, whose knowledge of Continental science outstripped his rivals', recalled that Alexander von Humboldt had already experimented with the rapid oxidisation of zinc in wet contact with silver, and he tracked down and translated the work of another German scientist, Wilhelm Ritter, who had shown that this reaction radiated out in a 'series of galvanic circles'.[39] With this advantage, the stage was set for Davy to demonstrate the reach of his experimental genius.

Davy worked as furiously as he had with the gas, not leaving Dowry Square for days on end and often sleeping no more than two or three hours a night. He needed to work at a frantic pace, and to conceive his experiments with maximum precision to work through all the variables that the voltaic pile offered in order to uncover its mechanism of action. He began by demonstrating that water released its component gases in the ratio that made up the original substance. Next, he substituted caustic potash for water and found that the same gases were released, but more quickly, demonstrating that some substances were better conductors than others. He then substituted charcoal for one of the metal strips, and found that gas was still released, but more slowly: charcoal, therefore, also conducted electricity, but either did so more slowly or absorbed the gas that was produced. He showed that the pile was inactive under water unless the water was saturated with oxygen; experimenting with other liquids, he found that muriatic acid was a better conductor than water. Using concentrated nitric acid, he built a pile of eighteen plates that was more powerful than Volta's own pile of seventy. By the end of the summer he had proved conclusively that the chemical changes produced by the pile were always the same, but that the effect was stronger with some substances than others. He compiled a table ranking dozens of substances by conductivity, and concluded that the most powerful pile of all was composed of copper plates separated by layers of cloth soaked sequentially in powerful acids and alkalis. He had effectively invented the cell battery.

Davy's breakneck dash for priority was reflected in the letters pages of *Nicholson's Journal*: from September 1800, every issue had a fresh bulletin from him, and by November it was clear to the scientific public that he had, at least in broad brush, unlocked the secrets of the pile's mysterious mechanism. In October he finally took a break from his experiments to travel with John King and Charles Danvers, a friend of Tom Poole, by boat from Bristol's Welsh Back Quay up the Severn estuary to the mouth of the river Wye, and from there through its steep wooded valleys to view Tintern Abbey, which Wordsworth's lines had turned into an icon of the ravaged sublime, under a full moon. His goal was, he wrote to Giddy, 'perfectly accomplished': they watched for three hours 'all the varieties of light and shade which a bright full moon and a blue sky could exhibit in this magnificent ruin'. He also confided to Giddy, now back in Tredrea, his first tentative summary of the theory to which his voltaic researches had led him: 'Galvanism I have found, by numerous experiments, to be a process purely chemical', with the surfaces of different substances 'having different degrees of chemical conducting power'. The chemical changes produced by the pile, such as the oxidisation of metals, 'are somehow the cause of the electrical effects it produces'.[40] Where Volta and those who followed him had established that the metals produced electricity, Davy had recognised that this electricity was in turn powering a chemical reaction that separated substances into their components just as heat did, building up acidic and alkaline residues at different poles.

This new world was indeed as strange as Priestley had predicted: electricity was ultimately chemistry, and chemistry electricity. It was proof positive that Lavoisier's chemistry, focused as it was on the material elements, could never describe the fundamental mechanisms of nature, because the properties of these elements were defined at a deeper level by the invisible forces that governed their interaction. It was confirmation of the path that Davy had started upon in the *Researches* by showing how nitrogen and oxygen could take a multitude of forms in different ratios, each with quite different properties from one another. Now, he had shone a light on mysteries that transcended Newton's universe of corpuscular motions, with its mechanical actions and reactions; the world he was revealing reflected far more closely Schelling's *Naturphilosophie*, where the visible world was merely the play of the forces and fluxes that underlay it. Davy's gaseous revelation that 'nothing exists but thoughts' might, it seemed, be more than a metaphor: the material world was an illusion generated by a deeper force, perhaps ultimately the ideas and impressions in the mind of its Creator.

The Pneumatic Institute had promised the world a miracle; when it had failed to emerge, the doubters and mockers had revelled in the spectacle. But they had made their judgement too soon: a momentous discovery had indeed emerged, though it was not the one that had been trumpeted. Lightning, it seemed, had struck twice.

*   *   *

By the end of 1800 Humphry Davy's name and reputation had spread throughout Britain's scientific community and beyond. His *Researches* had been reviewed widely and effusively, not only in the most important scientific and medical journals, but also in cultural and literary forums such as the *Monthly Review*. According to the thirty-page review in the *Annals of Medicine*, it demonstrated, 'in remarkable degree, the industry and genius of the author, whose enthusiasm for the discovery of truth has led him, with a perseverance almost without example, to undergo very great sufferings, and even to run the risk, on many occasions, of immediate death'.[41] His bold self-experiments were repeatedly highlighted in the reviews, and they became the hallmark of his developing persona. In the steely scientific context in which he had set them, they escaped the taint of subversion and self-indulgence that had attached to Beddoes and the poets: they seemed rather to embody the heroic devotion to knowledge that the sciences of the new century would demand.

The public had barely had time to digest the *Researches* before Davy's bulletins on the voltaic pile began to pepper the scientific journals: he had, in a matter of months, passed from an unknown laboratory superintendent to the most exciting chemist in the country, and an object of attention far beyond Bristol and the Pneumatic Institution. His researches were being widely repeated and duplicated. As early as March, on the basis of his first notes in *Nicholson's Journal*, the Askeian Society, a London philosophical forum that met to replicate novel experiments, had succeeded where others had failed in synthesising nitrous oxide and repeating the Bristol circle's dizzying intoxications: the lecturer William Allen 'had the idea of being carried violently upwards in a dark cavern with only a few glimmering lights'.[42] Now, many more were building voltaic piles and waiting for Davy's latest report, while others were beginning to wonder whether Davy might be lured from Hotwells to work under their auspices. Prominent among these was the geologist and artist Thomas Underwood, a proprietor of London's newly established Royal Institution who had, even before the *Researches* was published, proposed Davy as a possible chemistry lecturer to its founder and director, Count Rumford.

Rumford, who had been an American mercenary named Benjamin Thompson before being made a count of the Holy Roman Empire for military services to Prince-Elector Karl Theodor of Bavaria, was well known to both Beddoes and Davy: Beddoes had sent his famous paper on heat to Davy in Penzance before they had even met. His scientific fame rested on a grand experiment in which he had measured the heat generated by friction in boring a hole through a brass cannon and demonstrated that it had been produced simply through motion: like Davy's early demonstration using melting ice, it was designed to show that friction could be explained without resorting to Lavoisier's notion of caloric, which Rumford had long believed was 'merely a creature of the chemist's imagination'.[43] He proposed instead that heat was simply generated by motion, as the anvil heated under the hammer.

But Rumford was only a part-time scientific experimenter: his chief interest, which he had developed while working with the Bavarian army, was the application of scientific discoveries to daily life, and by extension the production of new inventions to help the poor. He had pioneered new designs for economical thermal devices such as chimneys and fireplaces, had invented the double boiler and the heat-efficient 'Rumford stove'; he had also, in parallel with Beddoes, campaigned for economies of food and the use of pressure cookers, disseminating a recipe based on barley and potatoes that became known as 'Rumford soup'. In 1798 he had relocated to Britain, on whose side he had fought during the American War of Independence, and had mobilised his aristocratic connections to form a Society for Bettering the Conditions of the Poor, with George III as its patron; the following year, with the support of his close friend Sir Joseph Banks, he had launched a new institution dedicated to matching 'new applications of science to the useful purposes of life'.[44] The Royal Institution was set up in a grand terrace off Piccadilly that had been remodelled to house a collection of practical machinery such as stoves, boilers, ventilators and kilns, and also a chemical laboratory and lecture room where experiments could be presented to the public.

Rumford may have shared Beddoes' goals, but his character could hardly have been more different, as their two Institutions indicated. While Beddoes had placed himself in modest style among the sick and drooping poor of the provinces, Rumford's premises were intended to establish him at the hub of London's elite society. He had begun his career in America by marrying a wealthy widow, and had proceeded to install himself at the royal court in Bavaria; now, his intention was to cultivate the highest circles in London. He was a man of notoriously easy morals, charming and seductive, and often shameless in his use of others. His charitable instincts towards the poor were genuine, and his connections with the aristocracy and high society allowed him to raise funds for them on a scale that Beddoes could never hope to match; but it was equally central to his interest in the Royal Institution that it would function as a public face for his personal empire.

In establishing the Royal Institution, as in all his endeavours, Rumford acted as a confirmed autocrat, and his insistence that its every initiative should bear his personal stamp was already creating difficulties. His current chemistry lecturer was Dr Thomas Garnett who, like Beddoes, was a former pupil of Joseph Black, a practising physician and an established lecturer, and his first course of lectures had proved a great success; but when Garnett had proceeded to publish the outline for a second course without consulting Rumford, he had suddenly found a committee set up to oversee him, and his pay frozen until the revised outline had received his director's seal of approval. When Rumford heard Underwood's report of a new star in the chemical firmament, his interest was immediately piqued; he became more interested still when he visited Edinburgh in September and found one of his friends, the chemist and artist Thomas Hope, clutching a copy of Davy's *Researches* and commending it as the finest work of chemistry he had read for many years.

Rumford discussed Davy with others, including a mutual friend in Bristol, James Thompson, who had participated in the nitrous oxide experiments, and by the end of the year the two were in contact through intermediaries. Davy's letters to his mother around Christmas began to include vague but tantalising hints of 'prospects of a very brilliant nature'.[45] By the beginning of 1801 he was confident enough to write to his old guardian, John Tonkin, and assure him that, after little more than a year away from Penzance, he was already certain that he had been justified in leaving his secure apprenticeship. 'The professors of the University of Edinburgh', he reported, mindful of Tonkin's ambition that he might one day take his medical degree there, had received his researches 'with great ardour', and 'I have received letters of thanks and of praise for my labours from some of the most respectable of the English philosophers'. 'I am sorry to be so much of an egotist', he signed off; but he wished Tonkin to know that his gamble had paid off, and that he was poised to leave the respectable dreams of a Penzance doctor far behind him.[46]

By the end of January 1801 he was finally prepared to confide more specific information to his mother. 'During the last three weeks', he revealed, 'I have been very much occupied by proposals of a very flattering nature', and he had accepted an offer to join 'the Royal Philosophical Institution established by Count Rumford and others of the aristocracy'. He was keen to stress that this represented a giant leap not just professionally but socially, and to assure her that he would no longer be a hostage to Beddoes' politics: 'you will all I dare say', he adds, 'be glad to see me getting among the royalists'.[47] Writing to Giddy, he expressed a similar sentiment, but with a rather different emphasis: 'Count Rumford professed that it will be kept distinct from party politics; I sincerely wish that this may be the case'.[48] He was already, perhaps, feeling a weariness with the battles not of his own making that, as long as he remained with Beddoes and the Pneumatic Institution, he would always be forced to fight.

Davy was able to assure Giddy, however, that Beddoes had reacted calmly and kindly to the news. Just as William Borlase had generously waived the terms of his apprenticeship in the face of Beddoes' offer in 1799, so Beddoes had 'with great liberality absolved me from my engagements'. Davy parted gracefully both from Beddoes and from Giddy, to whom he insisted that 'the nitrous oxide has evidently been of use' in therapy, and that Beddoes was 'proceeding in the execution of his great popular physiological work, which, if it equals the plan he holds out, ought to supersede every work of the kind'.[49] Yet as he left Bristol in early March to become Garnett's assistant as lecturer in chemistry – with the private assurance that he would be taking over the main role sooner rather than later – he was looking forward, and not back. His visit to London with Coleridge had convinced him that the metropolis was 'the grand theatre of intellectual activity', where 'society of the most refined kind offered its banquets to the mind, with such variety that satiety had no place in them'.[50]

8

A Victim to Experiments

The departure of Davy was not the end of the Pneumatic Institution: in fact, through the winter into 1801, the queues of patients grew longer. The nitrous oxide researches might have received more than their share of ridicule, but they had also brought the Institution to the national stage, and Davy's work had diffused an atmosphere of pioneering genius around Dowry Square. John King was proving a conscientious and sympathetic medical attendant, who restored patients' confidence that they were being treated for their own benefit rather than that of experiment. James Sadler's son took over from Davy as laboratory superintendent, freeing King to run the dispensary. Beddoes, meanwhile, found himself with the time to resume projects that the pneumatic researches had sidelined. He expanded his survey of syphilis and its treatment, produced another pamphlet on the management of consumption and scrofula, received his seasonal migration of private patients such as Gregory Watt, and began working intensively on the book that Davy had mentioned to Giddy, which was to be his definitive attempt to pass on his hoard of practical advice for the preservation of health and the prevention of disease.

In the wider world the political winter that had held Britain in the grip of war for so long seemed to be thawing. In April 1801 William Pitt, recognising that Ireland could not be dragooned by the British army indefinitely, proposed some limited steps towards political reform and Catholic emancipation; but George III, for whom any concession to Catholics was a dereliction of his vows as monarch, was unable to countenance the proposals, and Pitt was forced to resign. He was replaced by his childhood friend Henry Addington, a man weaker and more mediocre in every regard but one who lacked Pitt's vested interest in maintaining the war with which he had persevered for so long. For the first time in nearly a decade it seemed plausible to hope that Britain might be able to turn towards peace.

For those wishing to make a break with their revolutionary youth, the new face at the top offered an opportunity to reposition themselves with a modicum of dignity. Robert Southey, who had returned from Portugal and Spain with a profound conviction that Catholics could not be civilised, found himself taking George III's part, and thus in danger of being forced to regard Pitt's reforms as too liberal; but Addington's accession made it possible for him to ascribe the change to politics rather than to himself. 'It is not I who have turned around', he argued, as he prepared to make his home with Coleridge and Wordsworth in Keswick; 'I stand where I stood, looking at the rising sun, and now the sun has set behind me.'[1] But for others, the new leader was too little of a change, and had come too late to make a difference. Giddy, once more marshalling his county militia to put down ugly riots in Cornwall's mining towns, felt that British society had been fractured beyond repair. He blamed Pitt squarely for his 'most determined opposition to anything bearing the semblance of liberty', and concluded grimly that he had 'in all probability ruined the country'.[2]

Giddy, meanwhile, found himself in another difficult situation, this time one of great personal delicacy. Since convalescing for two months in Rodney Place the previous summer, he had begun to receive a stream of letters from Anna, typically informal and teasing in tone, but with increasingly serious and troubling undercurrents. To begin with, she presented her correspondence as a continuation of his therapy, and a distraction from the serious tasks that were weighing on him once more. 'Dr. B. says you would be well, if you were not shut up like a poor little bird in a cage', she told him. In Tredrea he was surrounded by anxieties and responsibilities; and, as Beddoes had cautioned him, 'the mind must not be screwed up to its highest pitch'. She wished that he were still in Clifton, 'that we might torment you, pull you by the coat – drag you where we please tear the red pocket handkerchief from your throat and a thousand nice little jobs of this kind'. Giddy had proposed, as was his wont with the opposite sex, that he and Anna treat one another as brother and sister; she was holding him to his bargain, and wished for him only to 'tell me you are pretty well, and that you forgive my impertinence'.[3]

Giddy kept up his end of the correspondence, with perhaps more duty than enthusiasm; but by the spring of 1801 Anna's tone had shifted, and the confidences were becoming more conspiratorial. She first tried, without much success, to draw him into advising her on private projects ('Pray does it take more talents to write a play or a novel?');[4] but now her enquiries became more pointed, wrapped in hypothetical scenarios that nudged ever closer to confession. Eventually she revealed that she had got herself into something of a 'scrape' with a man whom Giddy had met briefly the previous summer, and whom he had 'pronounced a very silly fellow'; but Anna had perceived 'great generosity of character and much sensitivity' in him, and had gradually discovered that he was trapped in a miserable marriage with a woman who 'has as bad a heart as his is good'.[5] Between anguished doubts as to whether she should be telling all this to Giddy, and frequent admonitions to burn her letters, she eventually confessed that this man had developed feelings for her, and she for him.

Giddy, though he been drawn out of his customary shell by Anna's spontaneity in Clifton, now found himself profoundly uncomfortable: he had no wish to know more of this than Beddoes knew. But Anna, it turned out, feeling that 'to deceive so very excellent a creature must require a very different heart from what I believe mine is', had already made a full confession to Beddoes. His response had surprised and disconcerted her: instead of the scene she had been dreading, he simply 'praised me for my sincerity and seemed to become fonder of me'. He seemed happy for her to continue seeing the man, though as it turned out 'business called him to town where he now is and has been for these last two or three months', and the affair, such as it had been, was no more. Relief, for Anna, mingled with an evident disappointment that Beddoes had regarded the business as trivial, and that the intrigue had been disposed of in such an uncomplicated manner. 'The Dr.', she concluded drily, 'has some singular peculiarities in constitution and character'.[6]

The exchange had begun on the pretext of asking Giddy's advice, but now the floodgates had opened, and Anna began to unravel a series of tales that added up to a chronicle of a frustrated and bitterly unhappy marriage. Davy had not been the first young man to whom she had found herself attracted, nor the last. Her husband was a wholly admirable man, whose qualities she could never hope to match – 'his views are larger, his soul nobler than mine' – and yet, despite herself, she could not help also finding his admirable qualities oppressive. He was tireless in his devotion to the sick and to the poor, to reforming politics and illuminating the world; but his exertions for the public good left meagre rations of love and attention for his wife. Though he was all goodness, she could not help having 'moments when I am selfish enough to wish, not that I should be equal to him, but that he should be humbled to my level'. These were frustrations that had remained unspoken, and had become unspeakable: Beddoes indulged her when she so demanded, but without comprehending why she did so. 'I used to be told I was a very modest woman', she concluded to Giddy; 'I have forfeited this title entirely by having revealed these things and still more by having done them. I have half a mind not to send this.'[7]

Giddy had little to offer but stiff advice about the institution of marriage, the inflexible nature of its duties and an oblique reference to sexual frustration, informing her that it was extremely common for couples to find obstacles 'to that complete happiness which young women rather fancy they can attain'.[8] Anna was prepared to concede that her vision of the institution of marriage may have been unrealistic: 'one thing I have always thought, and believe always shall think, is that I would rather be mistress to the man I love than wife, but since this is contrary to all custom, and cannot be, I must be contented with being wife in the usual way'.[9] The matter hung agonisingly between them until, over the summer of 1801, Beddoes suffered an infection that led to a severe lymphatic swelling in the chest, and Anna was struck by the terrifying prospect of life without him. 'He is the best husband and the best man in the world', she wrote to Giddy; 'I believe I love him better, for I was so unhappy and anxious when he was ill I did not know what to do'. Beddoes' breathing, always wheezy, had become painful and desperate to the point where he was struggling for air and, Anna confided, in such agony that he begged her 'in the most serious manner to let him put himself out of pain – he repeated this two or three times, saying he could bear it no longer'.[10]

Once more, Beddoes had left Anna feeling petty and selfish, and horribly guilty for attempting to blame her weakness on him. Her confessions to Giddy, so painfully wrung out of herself, had been rendered trivial by the intimation of her husband's mortality. But they were also superseded by much more welcome news: Anna was pregnant. In December she gave birth to their first child, a baby girl, who was named after her mother. It was, she recognised with hindsight, childlessness that had left her feeling bereft, neither wife nor mistress. Unlike her cleverer sister Maria, she was unable to join in the learned conversations on medicine and chemistry, politics and philosophy that buzzed constantly around her, and yet she had been without a domestic sphere beyond that offered by running a household that fluctuated between a busy office and an empty shell. Parenting by proxy for the 'dear little Lambtons' had given her great joy, and she had been very conscious of how 'Dr. B took more pains with those children, showed them more kindness and exerted himself in every respect for them more than I could have supposed it possible for him to have done for any two individuals'.[11] Little Anna made her, finally, more wife than mistress, and the letters to Giddy dried up.

* * *

Meanwhile, life in London was exceeding Davy's expectations. Count Rumford moved him into the Royal Institution building, and settled on him the agreed salary of 100 guineas. The room he was allocated was Garnett's, who had no idea that an assistant to his post had been appointed until his lodgings were commandeered; in ill-health and at loggerheads with Rumford, he resigned, making Davy, as promised, the senior lecturer in chemistry within weeks. He picked up work in his new Institution's laboratory exactly where he had left off in Hotwells, and within two days of his arrival was busy once more with zinc, silver and copper plates.

Outside the Royal Institution he found the glittering social world he had anticipated. He joined the earnest young philosophers of the Tepidarian Society for their meetings in Old Slaughter's Coffee-House on St Martin's Lane, leaving Coleridge hoping that his gregarious temperament 'would not bring too many idlers to harass and vex his mornings'.[12] Although Coleridge had removed himself to the Lake District, Davy took up with the circle to which he had been introduced on his first trip to London; but he was also mixing with society at its highest levels. While Rumford was the figurehead and public face of the Institution, much of the network that supported it was Sir Joseph Banks', which not only included the metropolitan ranks of gentlemen-scientists but reached deep into George III's court. Indeed, much of the Royal Institution's charitable funding for provision of the poor came not from wealthy Londoners but from Banks' friends in the landed gentry and aristocracy who were alarmed by the scarcity and riots that threatened their country estates. As part of the Institution, Davy found himself close to the Royal Society, to which he was promptly asked to deliver a paper on the voltaic pile and its galvanic combinations. He would soon receive the coveted title of fellow, at the age of just twenty-three.

Davy was particularly delighted, too, to find Tom Wedgwood as a neighbour. Tom's health since his return from the tropics had proved as unpredictable as ever by stabilising over the winter at Gunville, and he had now taken up residence in the London offices of Etruria, the family business, in St James's Square, a few minutes across Piccadilly from Davy. Although still weak, he was able to manage some laboratory work, and he collaborated with Davy on researches that would, with hindsight, assume great significance. Although the documentation is scanty, he seems to have succeeded in producing what he referred to as 'silver pictures', which, when officially discovered forty years later by William Fox Talbot and Louis Daguerre, would be known as photography.

This was a discovery that followed on from the intensive experiments of 1792 that had led to Tom's breakdown. His work on heat and light had caused him to notice that silver nitrate darkened when exposed to light; during the autumn of 1800 in Gunville he seems to have explored this effect further, experimenting with microscopes and coloured lenses. Now, in London, he moved his researches to the laboratory at the Royal Institution, where he and Davy could work alongside and assist one another. The fruits of this collaboration are recorded in a paper printed in the first and only issue of the *Journal of the Royal Institution* in 1802, entitled 'An Account of a Method of Copying Paintings upon Glass, and of Making Profiles, by the Agency of Light upon Nitrate of Silver'. The method, according to the title of the paper, was 'Invented by T. Wedgwood', and is described 'with Observations by H. Davy'.[13]

The paper is brisk and rather impersonal in style, lacking the flourishes that Davy tended to bring to descriptions of his own experiments, but within it the principles of silver nitrate photography are enumerated clearly enough. A sheet of paper or leather, if painted with silver nitrate, undergoes no change when it is kept in darkness, but as soon as it is exposed to light, 'it speedily changes colour'. The silver nitrate coating is least sensitive to red light, reacting more markedly to yellow and green, and most of all to blue or violet. Any image thrown upon the paper is etched onto the surface in precise detail, and by this method 'the outlines and shades of paintings on glass may be copied, or profiles of figures produced, by the agency of light'. The system can also be used to make magnified 'images of small objects, produced by means of the solar microscope'. Once silver nitrate has darkened in light, the process cannot be reversed, making images created by the technique 'in a high degree permanent'.

Yet the predominant tone of the paper is not one of discovery, but of frustration and defeat. The darkening of the silver nitrate is permanent, but the images are not. As they continue to be exposed to light, they darken further, and eventually the coating becomes completely black; the sheet can be kept in darkness, and carefully examined in dim light, but 'no attempts that have been made to prevent the uncoloured part of the copy or profile from being acted upon by the light have yet been successful'.[14] Tom had invented the photographic image, but he had not discovered a means of fixing it at the optimum point of development. Without this, the photosensitive properties of silver nitrate were a curiosity, no more than a means of giving a half-life to a shadow; interesting, but not significant enough to interrupt Davy's galvanic experiments for long.

As a result silver nitrate photography hid in plain sight for a generation. The *Journal* did not circulate widely, and after it folded was quickly forgotten. Those who were aware of it assumed, as Davy's reputation as an experimentalist reached towering proportions, that the problem was insoluble. Talbot recalled in 1839 that a fellow-scientist had been deterred from investigating it by Davy's account; Talbot himself was fortunate in having discovered a fixative before he read it.

Davy was exhilarated by the freedom of his new laboratory where, as he put it to his former lodging-mate John King, 'I am about a million times as much a being of my own volition as at Bristol'.[15] His other obligation, however, was to communicate his discoveries to the public, and for this his schedule was a demanding one. With the resignation of Dr Garnett, he was obliged to give three series of lectures over the spring and summer: he prepared two on galvanic phenomena, and one on pneumatic chemistry. Right from the opening lecture on 4 March 1800, when the surrounding streets were gridlocked with waiting carriages, they were a resounding success: in front of an audience, Davy found that he could harness the energy he had first channelled in declaiming his poetry to the surf of Mount's Bay. Rumford had feared he would prove too uncouth for a sophisticated urban crowd, but his passion, extravagance and confidence proved infectious, and as the courses progressed his audience mounted to five hundred, many of them ladies of fashion who whispered that 'those eyes were made for something besides poring over crucibles'.[16] He revealed a unique talent for infecting his audience with the thrill of discoveries snatched from the heat of the laboratory, and it would be lecturing as much as experiment that would establish him as the foremost chemist of his generation.

His final series of summer lectures culminated on 20 June 1801 with an account of respiration, after which nitrous oxide was offered to the audience. With Davy as master of ceremonies, setting the stage for public participation and expanding lyrically on the effects of the gas, the experiment reached the heights of antic performance that it had so often failed to generate outside the Pneumatic Institution. Several 'philosophers of eminence' breathed the gas, with 'truly wonderful' effects; Thomas Underwood, who had first proposed Davy to Rumford, 'experienced so much pleasure from breathing it that he lost all sense of everything else, and the breathing bag could only be taken from him at last by force'. The reports that appeared were as breathless and fantastical as anything that had emerged from Dowry Square; but now the pneumatic revellers struggling to make themselves understood were the metropolitan establishment. 'The irresistible tendency to muscular action produced by this gas', concluded the Royal Institution's reporter, 'was such as cannot be described. It must be witnessed to be conceived.'[17] Davy was ecstatic: 'there was respiration, nitrous oxide and unbounded applause', he wrote triumphantly to John King, and 'the voice of fame is still murmuring in my ears'.[18]

Yet it was to be the last time that Davy would experiment with the gas in public. It was also, and not coincidentally, the evening that generated the image by which the nitrous oxide researches would be most frequently recalled to posterity, and which appeared the next week in the window of James Gillray's Bond Street gallery. *Scientific Researches! New Discoveries in Pneumaticks!* was a classic production of Gillray's jaundiced eye and fevered pen that, rather in the manner of the symposium of 'The Pneumatic Revellers', contrasted the flatulent absurdities of the experimenters with the earnestness of their scientific intent. Davy is a hirsute, gnomish figure who sniggers as he brandishes the bellows, and there is perhaps some play with the idea of the great and good being lured to such indignities by a mocking rustic, but the thrust of the satire is essentially the ludicrous spectacle of the philosophers at play.

There were many angles to Gillray's satire, but assessing the accuracy of new scientific theories was not among them: his target was not so much the pneumatic researches as the Royal Institution crowd, and the absurd light that nitrous oxide had cast on them. The following year, for example, he would publish a similar image entitled *The Cow-Pock! or, The Wonderful Effects of the New Innoculation!*, depicting Edward Jenner in his clinic surrounded by a crowd of dumpy gentlewomen and gormless yokels, all staggered to find miniature cows exploding out of their arms and faces. His point was not that vaccination was a quack remedy, a folly or a scam: it was rather that its acceptance into mainstream medical practice did nothing to expunge the underlying grotesqueness of the proposition. Similarly, *Scientific Researches! New Discoveries in Pneumaticks!* was not an attack on pneumatic medicine so much as a sideways look at the potential of experiment to distort the decorous world of the scientific elite into gross indignity.

*   *   *

Scepticism about vaccination was by this point increasingly marginalised, and Beddoes' continued dissent was placing him in unusual company. Jenner's studies and programmes had now been replicated around the world, and those who held out against them were typically conservative figures such as the former surgeon-general of Jamaica Benjamin Moseley, who also continued to insist that slavery was a humanitarian intervention by which Africans were welcomed into the global economy. Jenner himself suspected that Beddoes' continued equivocation – first arguing against the phenomenon itself, then insisting that vaccination was only a traditional folk remedy dressed up in modern medical jargon – had its roots in professional jealousy. If so, Beddoes succeeded at last in overcoming it. Jenner's friends had mounted a petition to Addington and the Treasury asking that he should receive an honorarium for his years of work; in September 1801, at a meeting in Bath, Beddoes stood up and added a proposal for a national subscription in case the parliamentary grant should prove too miserly (it would come in 1802, and at the handsome level of £10,000). From this point on, he campaigned vigorously on Jenner's behalf. He had perhaps come to accept that, in the annals of preventive medicine, one name would ring out across the world from the west of England, and it would not be that of Thomas Beddoes.

Yet the Pneumatic Institution was ever more successful in its own right. From January to April of 1802 John King logged 678 new admissions; by the summer three hundred new patients a month were processing through the doors of Dowry Square. In subsequent decades, through the Chinese whispers of Davy's and Coleridge's biographers, the assumption would take hold that once Davy and the poets had left Beddoes' circle, the Institution lingered on only as a shadow of its former self. In fact, it flourished as never before. Particularly often repeated has been Joseph Cottle's claim that Beddoes was forced to offer money to patients to induce them to become guinea pigs for his experiments; but this is a distortion or misunderstanding of the practice adopted by the Institution of asking patients for a deposit of a couple of shillings – or, for the poorest, 'a knife, a thimble or a ribband suffices'[19] – as surety that they would complete their course of treatment. At the end of treatment the deposit was returned: this was not a bribe to lure unwilling subjects, but a strategy to avoid wasting the valuable resources of an overworked staff.

Other testimonies alongside John King's statistics point to a busy and successful practice. A visiting physician from Vienna, Dr Joseph Frank, made a tour of inspection of medical practices in Britain at this time; he was sufficiently impressed by reports of the Institution to make a detour to Bristol to visit it and, once he had arrived, to cancel his next appointments to stay and observe longer. He recorded that Beddoes 'presides over this Institution with considerable zeal'; but nothing made such an impression on him as his first encounter with the doctor in Rodney Place and its legendary library. 'The first words that he addressed to me', he remembered, 'were "Which Dr. Frank are you? For there are a great many of you."' Beddoes produced from under his arm a pile of books, all in German and all by authors named Frank, 'constantly asking as he turned them over "Is that you? Is that you?" ' He was delighted to discover that his visitor was the author of a tract on Brunonian medicine, which he settled down to discuss in a German that was as good as his English.[20]

As well as running a thriving clinic, Beddoes was also attempting to produce the *magnum opus* that Davy had announced to Giddy at the time of his departure. This had its origins in the proposal for a new journal, *Nature and Man*, that had been announced as part of the nitrous oxide researches in 1799, but Beddoes had now reconceived it as a book, with which he had been struggling throughout his serious illness of the previous summer. In 1802 it began to emerge as a monthly series of essays, and was eventually collected into a single volume the following year. Its title was *Hygëia*, and it was advertised as 'A Series of Essays on the Means of Avoiding Habitual Sickness and Premature Mortality': it was the fruit of the plan, first proposed in the introduction to his lectures five years previously, to forge a coalition between the medical profession and the general public that offered, as Beddoes had come to believe, the possibility of a healthy society that neither could achieve alone.

The target audience for the work was indicated by a further subtitle: 'On the Causes Affecting the Personal State of our Middling and Affluent Classes'. This was not primarily a work for physicians, or for the poor he had so often addressed in the past, but for the wealthy whom he had so frequently attacked as the root cause of the nation's sickness. He had not softened his views about them but he had decided, rather than haranguing them, to try to win them over to his cause. As he wrote to Giddy while in the thick of *Hygëia*'s composition, 'it is good to hinder rich people from being content to be the diseased supporters of a diseased population': all the while that they 'soak in inactive luxury', they are poisoning the blood of the body politic, wasting resources for which others are gasping, and forcing them to live under the oppression driven by their commercial desires.[21] Yet, as Beddoes now recognised, the wealthy were also the only class with sufficient leisure and cultural access to absorb and diffuse his advice effectively. Just as their economic power was disproportionate, so was the influence they wielded, and the mechanisms available to them to act. One rich man converted to his views could accomplish more than a hundred poor.

Hygëia's essays revisit many of Beddoes' most familiar themes – primers of anatomy and crisp pen-portraits of illnesses such as catarrh, scrofula and consumption – but the view is a wider one than he had taken before: a *tour d'horizon* of the hinterland of medicine, the management of the body and mind, and the conduct of life as a whole. We should know our own bodies, and those of us responsible for others, particularly the heads of households, should regard policing the health of those in their care as an essential part of that responsibility. Even the wealthy should eat simply and modestly, abstain from the indulgence of too much meat, the enervating luxury of tea, and particularly the slow poison of drunkenness, arch-waster of money and energy and scourge of social cohesion. The expensive private boarding schools in which the wealthy immure their offspring are injurious in every way: they deprive children of the love of a family unit, segregate the sexes at a delicate age with often unhappy results, sanction the brutality of corporal punishment, and constrict the natural and healthy impulse to exercise and run free. Hypochondria and hysteria, liver disease and gout are occupational diseases of the wealthy as much as typhus and miner's lung are of the poor: the difference is that the rich have the means to control and treat them. Living frugally and conscientiously is not, for the wealthy, simply a pious act for the benefit of the poor, but one profoundly in their own interest.

Beddoes was reaching out beyond the divisions that had excluded him from the mainstream of British society over the previous decade, and he was rewarded. *Hygëia* was a success, appealing strongly to the socially conscious paternalists who were now supporting Rumford and the Royal Institution, and whose sensibilities were coming to permeate ever more branches of the establishment. It brought Beddoes back into alignment with Coleridge, who was attempting to awaken the same sentiments in the same sectors of society (and who also conceived of himself as suffering from many of the disorders that Beddoes described, including gout and epilepsy): he praised *Hygëia* to Southey as 'a valuable and useful work'.[22] But it won its new readership at the expense of the pithy, gadfly style that marks out the best of Beddoes' prose: its essays roll on at length in sonorous and repetitive cadences, their advice wrapped in pious sentiments and homilies. Beddoes was conscious of this fault, and continued to work on a condensed text that never materialised; but his new style was well judged for the new times, as the allusive, scattershot manner of the old century gave way to the verbose *politesse* of the new.

Beddoes' new audience was one to which Davy, at the Royal Institution, was also tailoring his message. On 21 January 1802, in the introductory lecture to his first full course of chemistry lectures, he mesmerised the packed crowd, Coleridge among them, with a vision of the bright day that was about to dawn under the sun of science. In his conclusion he stressed that the new science would transform society, but not in the destructive ways that the previous generation had feared. 'We may look forward in confidence', he announced, 'to a state of society in which the different orders and classes of men will contribute more effectually to the support of each other than they have hitherto done'. Plans and projects for the amelioration of the lot of the poor were everywhere under way, and 'the most powerful and respected part of society, are daily growing more attentive to the realities of life', which must lead to them 'becoming the friends and protectors of the labouring part of the community'. But the poor would always be with us, and indeed provide a crucial function: 'the unequal division of property and labour, the difference of rank and condition among mankind, are the sources of power in its civilised life, its moving causes, and even its very soul'.[23]

This was a new chemistry for a new public, and Davy was explicit in distancing it from the rhetoric of revolution with which it, and by association he himself, had become entangled during the dark days of the 1790s. With the Peace of Amiens finally concluded on 25 April 1802, the nation was emerging blinking from the fog of war and wondering how it could have come so close to tearing itself apart. In the new consensus it was accepted that riots and starvation were scars on the body politic that must be healed, but that this could be accomplished without the surgery of revolution. Davy was giving scientific credence to this dispensation: society, like the new world of chemistry, was a flux defined by opposing forces, whose vital energies were drawn from the tensions between its elements. The attempt to equalise society was to be left behind with the old century; the wild gas had been tamed.

* * *

Beddoes may have altered his style to accommodate his new audience, but his message remained unchanged: he saw in science no justification for social inequality, and indeed glaring evidence to the contrary. *Hygëia*'s advice to the rich was only the first half of the *magnum opus* he had been gestating. The body politic that he witnessed daily in Hotwells was still hideously disfigured by a seemingly endless procession of the downtrodden and desperate, generated by a society deformed more than ever by the demands of industry, money and war. With the collapse of the pneumatic experiments, his hopes of a chemical revolution in treatment for them had been, if not abandoned, at least deferred. His final candid assessment to James Watt was that 'I have no confidence in my old speculations nor tenderness for them – but I hold it as a fact that the gases have salutary powers – and that in high degree'.[24] But his imperative as a physician, he now realised, was to treat the poor with what he had. He began referring to Dowry Square as the Preventive Medical Institution and, with a donation of £150 from Tom and Jos Wedgwood, opened a new surgery at Broad Quay, right in the heart of the wharves and slums of Bristol's docks. 'The distance from

Bristol to the Hotwells must, at all times, be fatiguing to certain invalids', he recognised, reflecting the extent to which his patients were no longer drawn from the private boarding houses of the spa but from the reeking bowels of the city.[25] The Institution was officially renamed, becoming the Medical Institution for the Benefit of the Sick and Drooping Poor.

It was a shift driven not by project or experiment but by popular demand, and patients streamed in to the new clinic 'by thousands in the year'.[26] By 1804 John King's ledger had expanded to over eight thousand cases, and Beddoes published a chapbook, nominally the rulebook for the new Institution but, like *Hygëia*, ranging widely beyond its stated theme. The 'peculiar end of the institution', he announced, was prevention: it would continue to treat desperate cases, but would also expand its efforts to 'check the canker of disease as soon as it fastens on the frame, and root it out'.[27] Beddoes' roll-call of the triumphs of preventive medicine, both traditional and modern, now presents the modern marvel of vaccination high in its firmament ('the effect of the cow-pock is truly wonderful') and credits it with the power, if fully implemented, to save forty thousand lives a year.[28] The symptoms of consumption, scrofula and smallpox are described in detail, with particular emphasis on the earliest signs to encourage home diagnosis and treatment. The occupational diseases of the poor are followed back to their roots in diet and drink, lack of fresh air and ventilation; and he is sanguine that the cumulative effect of his advice is already effecting a revolution in treatment. 'The Institution', he concludes, 'has certainly rendered cases of consumption less frequent' in Bristol; and 'by studying the earliest signs of the disorder, and endeavouring to correct the disposition to it, we may expect to ascertain a method of preventive treatment, which shall stand to consumption in the same relation in which the consumption stands to the small-pox'.[29]

From the beginning Beddoes' projections of chemical medicine had drawn their urgency from the quest to find an effective treatment for consumption; in pursuing his dream, he had wandered from his original goal, but now he was returning to it. The new project lacked the revolutionary sweep of the Pneumatic Institution, but it was its superior in the essential calculus of achieving the greatest good for the greatest number. Its rulebook looked back to the *Guide for Self-Preservation* he had written on his arrival in Bristol a decade before, and addressed the poor in the same tones: urgent, authoritative, acutely observed and tightly focused on communicating the thousands of hints and prescriptions he had assembled over the years in the crowded pharmacy of his mind. The lofty encouragements and sententious tone of *Hygëia* are dispensed with: Beddoes was still, as he ever had been, more comfortable writing for the poor than for the rich. Gone, too, are the speculations into which pneumatic therapy had forced him: he is no longer obliged to defend Brunonian theory against its critics, to argue for the validity of experiment, or to dispute the nature of evidence. He can speak directly to his patients, and on the evidence not of a few experimental subjects but of the thousands who have passed through his doors.

* * *

In June 1803 Beddoes and Anna had a son, Thomas Lovell, whom they called between themselves 'the young doctor'; he would indeed become a doctor and a poet, and his fame would in time eclipse his father's. In August, Anna went for the first time to Tredrea, where she had been invited by Giddy's sister Philippa to nurse and recuperate. She took the opportunity to solicit Giddy's advice and tuition to improve her mind, attempting for the first time to read Herodotus ('I find the style so agreeable that when I am more at leisure I will read him through')[30] and rediscovering Samuel Johnson's *Rasselas*, which she had found very melancholy as a teenager, although 'as my father and sister despised and laughed at the book, I had a confused idea that it was something very absurd'.[31] But as her visit extended into September, it became ever more apparent that the programme of improvement was a charade. She was in the throes of another of her 'scrapes', and this time the object of her passion was Giddy himself.

Though Anna loved her husband deeply, she could not escape the realisation that Giddy was everything that he was not: tall, fair, well-built, handsome, and a steadfast and reliable public figure. Beddoes himself had come increasingly to depend on Giddy's judicious and solid temperament to anchor him through the turbulent years that his friend, though constantly frustrated and compromised, had navigated with a wisdom and patience entirely beyond Beddoes' powers. These were qualities that Anna needed as profoundly as her husband did, and her need had become ever more acute for that which Beddoes could not give her. Giddy was, predictably, horrified by Anna's interest in him: the relationship for which she yearned was unthinkable to him, a sentiment he expressed with the axiom that two parallel lines could never meet. Back in Clifton, Anna vented her fury at the tone of his rejection: 'surely you need not consider so very mathematically . . . what you say to me'. She wished that she was still in Tredrea so that she could 'cut a long sheaf of your hair off . . . and then see if I could not arouse you from your provoking apathy'.[32] 'I am angry with you', she told him, 'for humbling my proud or rather vain hopes'.[33] Once again, she had torn the veil from her most intimate feelings, only to be humiliated by her own foolishness.

But Anna was also angry because she knew that, despite Giddy's frigid exterior, the attraction between them was more than a one-sided infatuation. From the time that they had met in Rodney Place, Anna had touched and charmed Giddy in ways that he was unable to express but equally unable to control. In November 1803 Giddy passed through Bristol on his way to Oxford, and stopped overnight at Rodney Place; a letter, copied in Giddy's hand from an original lost or destroyed by Anna, testifies to a moment of unguarded passion. In Anna's carefully transcribed words, 'your languid head rested upon my shoulder when I fearfully sought in your eyes what I trembled lest you should find in mine . . . when our eyes asked each others lips leave to meet they did meet and from that moment I slept no more – Oh I cannot I must not go on'.[34]

Giddy's diary entry for the date reads only: 'the evening of this day rather remarkable'.[35]

*   *   *

Beddoes had noticed Anna's state of excitement on her return from Tredrea, but apparently not the anger or rejection that lay beneath it: he wrote to Giddy to thank him for his hospitality, and observed that Anna 'has gone on in such a tide of pleasurable feeling that she will hardly know how to live at home'.[36] But Beddoes was himself preoccupied at this point, not with love but with mortality. His father had died earlier in 1803, after several months of paralysis that had rendered him as helpless as an infant and led Beddoes to 'consider his death a release, which I should desire for myself'.[37] It also prompted him to reassess his family and to recognise that, despite their frictions, his father had worked hard throughout his life to maintain difficult relations on an even keel, an example that had 'essentially contributed to the constant serenity I have enjoyed at home and which is little likely to be disturbed by any internal cause'. Even his wife's tremulous consummation of her passion for his best friend would not now disturb his equilibrium.

But the sight at the funeral of old family friends who were now 'aged or infirm, whom years ago I remember young or robust – this was the most melancholy part of the whole', he wrote to Giddy. It had, he confessed, brought on a sharp and oppressive bout of 'the Hamlet tone of mind'. Learning how to cope with this affliction – 'what Solomon calls the "vanity" and Darwin the "nihility" of all things' – was, he was coming to feel, 'the greatest point in education'. The goal of education was, after all, to expand our understanding, and to allow us to see the world from the grandest perspective, yet in doing so we cannot help but glimpse ourselves as smaller and less significant beings, and our strivings and achievements as transient and futile. This is a perspective that is salutary and necessary in its place, 'but if it should be habitual – I would have it occur but seldom'. Yet as Beddoes reached his mid-forties, it was becoming an ever more frequent visitor, and bringing with it a suffocating sense of his vast learning turning in on itself: 'your mind perpetually changes into a huge vault in which the world and everything it contains lies entombed'. It was made more oppressive, too, by the fear that this would eventually be the fate of the young minds he was bringing into the world: 'I would not have a child be a living sepulchre of nature'. 'If you know any secret for this purpose, pray let me have it in a legible hand', he demanded of Giddy, noting that 'I see that you contemplative people, whatever other delights you may enjoy, are in general worse off than the active in this respect'.[38]

His father's passing capped a series of deaths, and presaged several more. The previous year Erasmus Darwin, seventy but seemingly as robust as ever, had died abruptly after being seized with a shivering fit; in June 1803 William Reynolds had finally succumbed to the long illness that had frequently sent Beddoes back to Ketley on mercy dashes over previous years. The generation who had taught and nurtured him were fading, leaving him to stand as the elder statesman; but, more tragically, the younger generation that he had taught and nurtured were falling too. In 1804 Gregory Watt finally succumbed to the consumption that Beddoes had battled for so many winters: however many lives he had saved from the condition in the Bristol slums, Beddoes was keenly aware that Gregory was the second of James Watt's children whose lives he had fought for and lost. Davy, for whom Gregory had been the *entrée* into Beddoes' world, and with whom he was in the middle of a warm correspondence, was distraught. 'He ought not to have died', he wrote to a friend when the news broke; 'I could not persuade myself that he would die, and until the very moment when I was assured of his fate, I would not believe he was in any danger'.[39]

Worse was to come. In April 1804 Coleridge, opium-haunted and despairing at the collapse of his marriage and his disastrous obsession with the Wordsworths' friend Sara Hutchinson, had exiled himself to Malta; his friends feared that he was a meteor burned out, never to return. He had asked Tom Wedgwood to accompany him, but Tom was too frail to travel: over the last year he had sunk into a state of almost permanent collapse, racked with internal pains of mysterious origin and taking ever larger doses of opium in his attempt to dull them. In July 1805, aged thirty-four, he was found immobile in his bed at Gunville. The doctor pronounced him dying, and he 'continued in that state, his head quietly reposing on his arm, till seven in the evening, when he expired without seeming to have suffered the least pain'.[40]

It was an ambiguous death, a wrenching loss and a merciful release in equal measure. 'It is difficult to know in what light to consider it', wrote Sydney Smith, a Dorset neighbour, to Jos. 'It is painful to lose such a man, but who would have wished to preserve him at such a price of misery and pain?'[41] Tom had suffered his undiagnosed and incurable agonies throughout his adult life, his researches snatched in moments of lucidity interspersed by crushing bouts of illness. His death was a moment of punctuation that marked, among much else, the eclipse of the Pneumatic Institution and its dreams of transforming medicine. The project would not have been possible without his funding; and nor might its experiments, without his intense cultivation of the language of feeling and sensation, have followed the path they did. Yet he had remained a 'Hotwells case', a patient for whom all other possibilities had been exhausted, and perhaps one who had never had any hope at all. From beginning to end his condition had remained a mystery, making it impossible to know whether the pneumatic therapy had failed, or whether it had ever stood a chance. Like his silver pictures, Tom's life had had its moments of dazzling clarity, but he had never found the secret of fixing them, and stopping them from fading inexorably to black.

*   *   *

Yet for Beddoes life went on, and in many ways more happily and prosperously than ever before. From those who could afford it, he was now commanding fees at the level that Darwin had, and being summoned to attend patients as far away as Yorkshire. He had invested much of his father's estate, under Giddy's super-vision, in American stocks; Giddy was extremely solicitous to warn him that the business was not without risk, but Beddoes' optimism was irrepressible, and he put Giddy's anxieties down to hypochondria. 'I wish I could persuade you to dismiss your troubles on account of my money', he told him;[42] 'the sight of so many higher, better, abler people degraded, impoverished, destroyed must reconcile me to misfortunes, if they happen, or where is the use in history?'[43] His stocks rose to over 8,000 dollars in value and he found himself, after so many years of scrabbling for subscriptions of 200 or 300 pounds, with similar sums of his own comfortably to spare.

The Preventive Medical Institution, too, was developing a reputation far beyond Bristol. By 1806 several of Beddoes' fellow-physicians had attempted to persuade him to move to London, leaving his clinic in the more than capable hands of John King and his new assistant, John Stock. Beddoes consid-ered the move, and seems even to have made some preparations for it, before a recurrence of his illness made it impossible. The prognosis was not good. His perennial wheezing breath had progressed into a dropsy on the chest, and his lungs oozed constantly with fluid; his liver, too, was showing signs of strain and intermittent failure. His little fat frame, which he had propelled up and down the cliff paths from Clifton to Hotwells every day and rattled endlessly over rough country tracks in stagecoach and carriage, with always another project, pamphlet or patient to attend to, was finally wearing out.

Anna was solicitous for her husband and devoted to his care, but at the same time unable to banish the thought that his death would leave her and Giddy free to marry. The latter remained as close to Beddoes as ever, their correspondence more frequent than at any time since the early days, Beddoes overseeing his health as he oversaw the doctor's investments. In May 1804, when Giddy was elected to Parliament as the member for Bodmin, Beddoes was delighted: 'I am very glad you are in Parliament – as I think it is the only chance for saving the country – it is a step to being Prime Minister which would be our certain salvation'.[44] Beddoes had not been so sanguine about politics for twenty years, nor had his hopes become any more realistic in the interval. He advised Giddy to be vigilant about his health in London, and to prepare for parliamentary sessions by sponging his body down with strong brine.

With Giddy's new status, the question of marriage became inevitable, though it remained an institution of which he had many terrors. In 1804 his sister Philippa had announced her intention to marry a mathematician named John Guillemard; Giddy had protested violently, even attempting to enlist Beddoes to certify the man a lunatic, but Philippa had faced down his protests, and the marriage had turned out to be a perfectly happy one. Losing the sister he had lived with all his life had left him vulnerable and alone, but the prospect of Anna seeking to fill the vacuum after Beddoes' death consumed him with dread. In August 1807 he closed off the possibility by proposing to an heiress named Mary Ann Gilbert whom he knew, though not well, from her family visits to Cornwall. At Christmas, when he next visited Clifton, Anna was in shock, and in private argued bitterly with him; Giddy, fraternally, assured her that it was for the best. Resignedly, she made her peace with him: 'Yes, you are right, we shall both be happier when all is settled'.[45] Giddy was consumed with misgivings before the wedding, and it was Anna who insisted to him that he must go through with it. During the course of the long and scrupulously formal marriage that followed, Giddy continued to keep his diary meticulously, but never mentioned his wife in it.

* * *

Beddoes had addressed his final statements to both the rich and the poor, but there was one last constituency that he needed to engage to take forward his reforms: the medical profession itself. The occasion for his message to them was a proposal aired by Sir Joseph Banks to take statutory measures to improve the quality of medical practice by controlling the profession's membership, regulating its conduct and 'suppressing, or at least restricting, empirical practice'.[46] Various regulations had been suggested – practising physicians should have a lower age limit of twenty-four, and degrees in medicine should be compulsory – but Beddoes had proposals that were considerably more far-reaching. His *Letter to the Right Honourable Sir Joseph Banks on the Causes and Removal of the Prevailing Discontents, Imperfections and Abuses in Medicine* (1808) was a submission that ran to nearly two hundred pages, and set out a programme of reform that would resonate far into the new century.

Beddoes concurred with the proposal of tightening controls on doctors: he had, after all, long dreamed of banishing 'unauthorised intruders into medical practice', and stopping the 'herd of quacks and itinerants dealing out their poison with unsparing hand'.[47] But this represented only the cleaning of the stables that was necessary before the serious business of reform could properly begin. 'The epithets romantic, visionary, extravagant or the like will be bestowed upon what follows',[48] he warned; but the world was changing so fast that a medical profession fit for the new century would in many ways be transformed beyond recognition. Most crucially, it must find ways to pool its knowledge and operate as a single unit in ways thus far never conceived. 'Why should not', Beddoes asks, 'reports be transmitted at fixed periods from all the hospitals and medical charities in the kingdom to a central board?'[49] The medical profession could no longer depend on a thousand doctors, each holding thousands of facts in their heads, but unable to compare their store with their colleagues. 'To lose a single fact is to lose many lives', he stressed; and yet, under the current system – or lack of one – 'ten thousand, perhaps, are lost for one that is preserved'. The medical profession of the future must become 'a national bank of medical wealth, where each individual practitioner may deposit his grains of knowledge, and draw out, in return, the stock accumulated by all his brethren'.[50]

Once physicians can conceive of themselves as such a unit, an army fighting disease, further proposals naturally follow. A unified profession will finally be able to quantify the health of the nation, to record the prevalence of different diseases, and to note their increase and decrease with different cures. 'No corpse ought to be allowed to be buried without the name of the fatal disorder from a medical attendant': national mortality statistics will finally become a matter of record, rather than anecdote and extrapolation from parish registers, and doctors will be able to know, rather than guess, the effectiveness of their preventions and cures.[51] The central board, or national bank of health, should have 'an office for proving popular as well as secret remedies', where the contents and utility of patent and herbal medicines can be properly assessed: seals of approval will be granted to those proved effective, and those that are not will be exposed as frauds.

Romantic, extravagant and visionary these proposals may have been in 1808, but Beddoes' last project was his most clear-sighted and prophetic. Here, in distilled form, was the programme that would transform medicine in the century to come. Ten years previously Beddoes had been on the losing side of a divisive and failed revolution: the future that he had envisaged so clearly, and for which he had agitated so enthusiastically, had been buried by events. Now he was on the crest of a breaking wave. In his new vision, science, medicine and politics were no longer in conflict but flowing in the same direction, borne on a tide of reform that, though he might not live to see it, would carry the next generation along in its sweep.

Yet there was one conspicuous casualty of the failed revolution: his familiar insistence that medicine would be transformed by pneumatics, or indeed by chemical medicine in general. The *Letter to Banks* is at pains to stress that newfangled and elaborate treatments are not necessary to his proposal: indeed, the whole 'art of physic may be acquired in three days by a person of the most moderate capacity'. At his Institution in Bristol, where he treats every disease that can be imagined, he has recourse to no more than 'a little vitriolic acid for the night-sweats, chalk-mixture for the bowels, poppy-syrup, or that favourite nostrum, the black drop, or what you please of the like, for the anodyne'.[52] Beddoes has learned from bitter experience that his claims of a chemical revolution have made him a hostage to unpredictable experimentation; but he cannot refrain from insisting that theories must still be advanced, and experiments diligently pursued. Such researches still offer the best hope of discovering new cures; but whatever their outcome, the pursuit of knowledge demands them for its own sake. 'The contemplation of organic and intellectual nature is a flower that blooms aloft in the purest fields of aether', he insists. Doctors cannot be restrained from philosophical enquiry and, if they discover the unexpected, they cannot ignore it: 'as well might one expect the astronomer to overlook eclipses and comets'.[53] He makes no claims for the value of factitious airs in the medicine of the future, but neither will he apologise for having pursued his researches wherever they happened to lead. Like a comet, after all, they may return, and reveal their secrets more fully to future generations.

* * *

While he was planning the future of medicine, Beddoes' interests again turned full circle, and he took up once more the geology that he had abandoned for pneumatic therapy. Through the exceptionally long, hot summer of 1808 he assembled a large collection of mineral samples, and experimented with pounding and heating them. He may have been testing new materials for road-surfacing: like Darwin with his designs for a hydraulic carriage, a life spent travelling up and down unpaved rural tracks had given him plenty of opportunity to contemplate improvements.

Beddoes and Anna had a second son in 1805, and in the autumn of 1808 they were joined by a second daughter. But as the cold and dark of winter closed in, Beddoes sank into a decline. He soldiered on through November, keeping an appointment in Wales, but by December he was weak and losing weight, taking soup and wine by the teaspoon and unable to leave his bed. With his world shrunk down to his bed, he became dependent on Anna; and once his conversation was humbled to her level, as she had once wished it might be, they slipped into the intimacy that had been so elusive during his busy daily round. 'In illness Dr. B. appears to advantage in many respects', she wrote to Giddy; 'his mind by being weakened dwells more upon individuals than upon general objects', and 'he expresses much more feeling and has less appearance of selfishness than when his mind is occupied upon larger objects'.[54] Beddoes, however, was still working. 'I have been under the necessity of investigating my own disease and its remedy', he wrote in a short note to Giddy: his condition, whatever else it might be, was a boon to self-experiment that should not be wasted.[55] The greatest relief he found was from bleeding by leeches: 'they had not been on five minutes, before the pain went, and has hardly been felt since'.[56]

In his last days he thought of Davy, who had recently made good all his grand claims about the new chemistry by using electrolysis to isolate seven previously unknown elements, which he named potassium, sodium, calcium, strontium, barium, magnesium and aluminium. It was a triumph of discovery never equalled before or since. In Lavoisier's system, the alkaline earths of these elements, such as potash and soda, had been taken to be their elemental forms; once again, Davy had shown his ability to make nature visible at a deeper level than anyone had previously suspected possible. 'Davy has just solved one of the greatest problems of chemistry by decomposing the fixed alkalis', Beddoes marvelled in November, taking a collegiate pride in the success, and smiling, 'on we go, deciphering the world'.[57] But by 22 December, when he wrote to Davy to congratulate him, the pitiless contrast between his former assistant's glory and his own state, diminished further by pain and sickness, gave his

Hamlet tone of mind full rein. 'I am', he confided to Davy, 'like one who has scattered abroad the *avena fatua* of knowledge, from which neither branch nor blossom nor fruit has resulted'. Despite the presence of Anna and John King at his bedside, he felt alone, 'in need of the consolation of a friend'.[58]

By Christmas Eve his breath had become desperately short, and he sank to the flickering limits of consciousness. Anna wrote a panicked note to Giddy, asking him to 'come and stay with our poor Dr. – he is influenced by you – I have not the power I would over his mind to make him take what is proper for him – the Dr. says it will be giving you a vast deal of trouble &c – but I know you too well to believe you will mind this now'.[59] But it was too late, and by the evening Beddoes had slipped away.

To his friends, Beddoes' death was a tragedy that, along with Tom Wedgwood's, brought the curtain down abruptly on a social circle that had felt like an enduring fixture in an ever changing world. 'Alas! Poor Beddoes is dead', Davy wrote to Coleridge. He had guessed the turn of events from the Hamlet letter of two days before, where Beddoes had poured out his soul in a manner 'full of affection and new feeling'; but he also felt the irony of Beddoes' passing at a moment when his projects and visions were poised finally to succeed: 'he is gone at the moment when his mind was purified and exalted for noble affections and great works'.[60] Coleridge, who had finally resolved to put himself under Beddoes' care for his opium addiction ('O God! Let me bare my whole heart to Dr. B.'),[61] received the news as a 'bodily blow' that provoked a 'long and convulsive weeping'. 'Beddoes' departure', he wrote, 'has taken more hope out of my life than any other event except perhaps T. Wedgwood's'. His addiction was still far too wrapped in pain and shame for him to confess it to the world, but 'Dr. B. was good and beneficent to all men, but to me he had always been kind and affectionate and latterly I had become attached to him by a personal tenderness'. He confessed to Davy that 'there are two things which I exceedingly wished . . . to have written the life and prepared the psychological remains of my revered friend and benefactor, T.W., and to have been intrusted with the biography of Dr. B.'.[62] These were both projects that would have transformed their subjects' legacies; but they were destined, like so many others before and after, to remain as glittering, and gradually fading, artefacts of Coleridge's imagination.

It was Davies Giddy, reliable as ever, who took on the burdens that Beddoes had left behind. He had agreed to act as guardian to his friend's children in the event of his death, and took care of wrapping up his business arrangements. The legendary library was auctioned at Sotheby's, which catalogued six thousand items; Rodney Place was sold, and Anna and the children ushered across the Irish Sea to Edgeworthtown. There was no family plot in Shropshire for Beddoes' body to be returned to: he was buried in the Strangers' Graveyard outside Clifton, alongside the multitude of Hotwells cases, those he had cared for, and failed to save, among them. Anna supported herself on Giddy's shoulder at the funeral, contemplating perhaps that, had Beddoes died a few months earlier, they might now be standing beside one another not in the grave-yard but in the church. Giddy would keep his word, and remain the guardian of her children for decades to come; they would be parents together, but he was the husband of another.

In the public sphere, Beddoes' death was taken as emblematic of the passing of an era. He was a man who had been defined by the great divide that had opened in the nation's soul: to some he had struggled nobly for enlight-enment and liberty, to others he had wasted his energies and talents in a doomed obsession with novelty. In either case, he had struggled in vain against the tide of history. The *Gentleman's Magazine* of February 1809 ran two extensive notices which showed that the polarised forces of revolution and reaction, though it was now the custom to situate them in the past, had not been entirely neutered by the new century. The first, by a Dr Crane, paid tribute to Beddoes in Darwinian couplets that were intended to transport the reader back to the exhilarating free spirit of enquiry that had pervaded the early 1790s:

> True Genius kindles fires, whose piercing light
> Reveals all Nature's secrets to the sight.

Yet even Crane was politely noncommittal about the Pneumatic Institution. Beddoes' most ambitious and notorious project was not mentioned by name, but disposed of in a glancing reference to a mind that, 'ever on the wing',

> Would not permit him to complete his schemes
> Which, hence, appear'd like visionary dreams.

Crane was too diffident even to attempt to arbitrate in the debate about Beddoes' ultimate legacy, concluding only that

> To Time the great decision will belong
> To prove both who is right and who is wrong.

The second notice was a prose obituary whose anonymous author, under the ironic byline of 'Amicus', had no such fastidiousness about the rush to judgement. In all the sentimental outpourings since Beddoes' death, he began, none had mentioned that he had 'a direct tendency to atheism', and indeed 'never scrupled to avow his contempt for Christianity', a sin that his death was not enough to set aside. Furthermore, in medicine 'he was such a theorist' that his sentimental obituarists should consider honestly 'whether as a whole he has done more good than harm as a physician'. He had spent his career taking risks, and usually on the basis of false theories and poor judgements; he had been as reckless with his patients as with himself, and 'after many around him had, probably, suffered, he at length fell himself a victim to experiments'. All in all, Amicus concluded, 'there seems something like what the common people call a judgement, in the dissolution of this great empiric of Bristol. He dies of experiments tried upon himself, to tell the world with an awful voice of admonition, that they who overstep the bounds of medical practice ought to be regarded with distrust.'

The claim that Beddoes had died of his own experiments travelled rapidly, and proved hard to root out. Before the end of December the local press had begun to repeat it, and by January 1809 the *Bristol Mirror* felt obliged to point out that the allusions to 'an experiment tried upon himself' were a 'shameful calumny'. It was a charge that had been proved wrong, and in the only way possible: by experiment. Beddoes' body was dissected by his assistants, John King and John Stock, his closest friends and the last in his lineage of scientific sons of genius. Their autopsy showed a congenital heart defect and a left lung that had clearly collapsed many years previously, the right lung massively distended in compensation. Yet, as the obituaries demonstrated, the appeal of poetic justice was more potent than the facts. Autopsies were the terror of the poor, who shunned hospitals for fear that doctors would experiment on their remains; while many had equally shunned the nitrous oxide treatment through fear of experiment, Beddoes had offered his wheezing body up for both. It was a fine irony that his final experiment, carried out on his own instructions, had revealed his fatal illness to have been natural and long-standing, the product of a handicap against which he had long struggled to his physical limit, and unconnected to the experiments for which he had been renowned and reviled.

EPILOGUE

The Last Days of a Philosopher

For the members of the Bristol circle who survived the cull of the early 1800s, it became ever harder to imagine that they had once shared common cause. Robert Southey was perhaps the most conspicuous example: if it was the case that he had remained true to the principles of his youth, as he claimed, then the world around him must have shifted on its axis. Within a decade he had become ubiquitous in the world of letters, and in 1813 he was appointed poet laureate, a post he would hold for the next thirty years. As a mouthpiece for the nation, his most characteristic posture was one of alarm at the threat of Jacobinism and subversion, and of aggression against those who advocated any form of dissent. In 1817 he used his position to urge the prime minister, Lord Liverpool, to adopt looser definitions and harsher sentences for seditious libel; the ensuing legislation succeeded in intimidating the new generation's champions of liberty such as William Cobbett into exile. After the Peterloo Massacre, he blamed the casualties on the revolutionary rabble, and defended the troops who had fired into the unarmed crowd.

Yet his memories of the Pneumatic Institution grew fonder with age. In the preface to the first edition of his collected poems, he recalled his year at Westbury as the happiest of his life – 'I have never, before or since, produced so much poetry in the same space of time' – a happiness that was vividly encapsulated in his first experiences with nitrous oxide. 'I was also then in the habit', he recalled, 'of the most frequent and familiar intercourse with Davy, then in the flower and freshness of his youth. We were within an easy walk of one another, over some of the most beautiful ground in that beautiful part of England. When

I went to the Pneumatic Institution, he had to tell me of some new experiment or discovery, and of the views which it opened for him'; and 'the bag of nitrous oxide with which he generally regaled me upon my visit to him, was not required for raising my spirits'.[1] Southey was distancing himself from the notorious experiments, but with the passage of time he was also able to recognise a dimension of the experience that had been more easily visible to those outside the circle than within it: the spirit of friendship had been as potent a trigger for their exalted moods and feelings as the gas they inhaled. When, after the first flush of the summer of 1799, the gas had ceased to produce for him its customary sensations of pleasure, the cause had not simply been the chemistry of nitrous oxide, but the dissipation of the atmosphere around it.

*   *   *

Davy, too, as he neared the end of his life, found himself turning to elegiac recollections of the same period. 'How well I remember that delightful season', he wrote in 1828, 'when, full of power, I sought for power in others; and power was sympathy, and sympathy power': it was a time when 'the great, of other ages and distant places, were made, by the force of the imagination, my companions and my friends'.[2] His life had indeed been, on the face of it, a formidable demonstration of the power of the imagination to shape reality. The teenager who had scribbled 'Davy and Newton' in his notebook had within five years overwritten Newton's vision of matter in motion with one of higher powers by which matter was constrained and transformed; by 1820 he occupied Newton's seat as president of the Royal Society, taking over the reins of patronage and power that Sir Joseph Banks had held for over forty years.

But, like Beddoes, Davy was only in his forties when his health began to fail. He, too, developed a heart condition, which in his case manifested itself in bouts of sweating and racing heartbeat, succeeded ever more frequently by palsies that numbed his right leg, and sometimes spread alarmingly up and down his entire right side. By 1828 he was travelling for his health, mostly in the eastern Alps – Austria, Italy, Slovenia – where the air was cool and pure and, surrounded by the high romantic scenery of waterfalls and precipices, he could indulge the childhood passion for fishing that he had never lost. He was accompanied by his godson John Tobin, the son of James Webb Tobin who had playfully cuffed Davy under the influence of nitrous oxide in 1799, and the presence of his young companion made it easy for Davy's thoughts to drift back to those days in Dowry Square. The bright and pendulous full moon, he recorded in his notebooks, filled his mind with the poetic reveries 'of my early youth, thirty-two years ago, when I was eighteen, and versified then in the *Annual Anthology 1799*'.[3]

Between feverish attacks that would leave him bedridden for days or even weeks, he worked with his customary energy on what would be his final two scientific papers. One, on the action of volcanoes, edged delicately towards the view of Hutton and Beddoes that they must be the emanations of a central fire deep in the earth's core; the other, on the physiology of the electric eel, continued his lifelong search for the principle that linked electricity to the animating forces of life. In tandem with both, in a notebook propped on his chest as he lay in bed, he worked intently on a visionary summation of his life's work entitled *Consolations in Travel, or, The Last Days of a Philosopher*, in which many of his early passions, dormant through his mature career, erupted once more to illuminate his mind.

Davy's final testament takes the form of a series of poetic dialogues with mysterious figures, elaborated from his dreams and restaged in spectacular settings under full moons, in the ruins of the Colosseum, on the slopes of Vesuvius, or against Alpine peaks and lakes. As they play out, the memories of a lifetime swirl together with dreams and visions, and with an interior debate in which aspects of the author, his heroes and his adversaries attempt to reach conclusions about the nature of progress, the spirit of the age and the future of humanity. In the Colosseum, where Davy finds himself wandering alone as night falls, the spirit of Genius materialises to carry him into the future, where he meets a race of demigods possessed of the genius that 'a few superior minds' have nurtured and brought forth.[4] In the fifth dialogue, 'The Chemical Philosopher', he meets the spirit of the Unknown, who reveals to him that chemistry is the art that offers humanity a microcosm of the workings of the Divine Intelligence, and the power to claim dominion over space and time. The overarching vision traces the journey of a vital spark, manifested and incarnated in the genius of a few individuals: 'most illustrious names' who 'were little valued at the times when they were produced', but who embody a force that will, as society comes under their enlightened thrall, lead the race towards its immortal destiny.[5]

Consolations is a vision of cosmic grandeur, but its narrative is recognisably Davy's own: that of the youth seized by the spirit of genius and directed towards the mastery of chemistry, through which he is led to the peaks of destiny. Yet its tone mixes triumph with a curious defensiveness: its clearly autobiographical subject is not a hero but an underdog and martyr, and its soaring visions are undercut with stabs of self-pity and bitterness at an ungrateful world. Davy had scaled the pinnacles of fame, but the view from the summit was not the sublime panorama he had anticipated. His resounding successes had been interspersed with less public but equally resounding failures, and with personal rifts and jealousies that had soured his view of life. His restless ambition had hurried him past the moments that, in his final months, he now recognised as the true peaks of his experience.

From the first, his landmark achievements had been shadowed by frustrations and disappointments. The Royal Institution, with its remit of developing practical applications for the benefit of the poor, had set him almost immediately to investigate the science of tanning, in search of improvements for the malodorous and unhealthy processes that had been in use for centuries. In analysing the tanning agents in current use, he had made pioneering discoveries – becoming the first, for example, to reveal the presence of tannin in tea – but had been unable to produce a chemical compound that was as effective as the traditional solutions of oak bark and gall used for generations by the likes of Beddoes' father and Tom Poole. As Beddoes had discovered, the chemist could readily produce substances not found in nature, but to find practical applications for them was harder than it appeared from the laboratory.

It was a procession of such failures that made Davy's design for the miner's lamp, when he unveiled it in 1815, an iconic moment in the burgeoning industrial revolution: by demonstrating in the laboratory that gauze would prevent a flame from igniting deep pockets of fire-damp, he had produced a much-needed advertisement for science's ability to transform modern life. Davy was borne aloft on a torrent of public praise, much of it orchestrated by himself: he received public acclaim from miners for 'this great and unrivalled discovery for preserving the lives of our fellow-creatures',[6] and was presented with a commemorative plate in Newcastle by the earl of Durham, whom he had first met as the older of the two Lambton boys on his arrival at Clifton. But even this moment of triumph was shaded with discontent. The chorus of praise was unable to silence doubts over Davy's claim to priority: accusations of plagiarism were levelled, which he was subsequently accused of using his position of power in the Royal Society to stifle. It rapidly became clear, too, that although the Davy lamp in theory made mining safer, in practice it enabled mineowners to send their labourers into deeper and more dangerous mines. The new chemistry, as Davy had proclaimed in his first lectures at the Royal Institution, was predicated on the inequalities of society.

The intense and passionate friendships that had defined and enriched Davy's youth in Clifton were also to become scarcer and more fraught as he rose towards the social stratosphere. Once installed in Piccadilly, he rarely returned to Bristol: with the death of Tom Wedgwood and Gregory Watt, and the departure of Coleridge to the Lake District, he had ever less reason to do so. In their place, he found a London society that became more formal and sterile as he ascended through it. In 1812 he was knighted, becoming the first man since Newton to be so honoured for his scientific work; three days later he married the widow of a baronet and, though his humble beginnings would always mark him as an *arriviste*, assumed his place in aristocratic society. The marriage was childless and unhappy: Lady Davy, as Sir Humphry always referred to his wife, had no passion for science, and drew him into a refined and cloistered world of country seats and London seasons where his friendships and passions, with the exception of fishing, were not shared.

In 1820 he was elected president of the Royal Society, the highest honour in British science; but his tenure was unhappy almost from the start. His youthful confidence had hardened into a public persona that was arrogant, brittle and defensive, demanding deference to his genius while showing an acute sensitivity to the mutterings that science had done more for him than he for it. In his attempts to remake Sir Joseph Banks' patrician network in a practical, meritocratic and industrious mould appropriate to the new century, he made enemies and flew into rages that brought on red-faced apoplectic attacks. At the helm of the scientific establishment, he had all the power he had ever dreamed of; but he had never dreamed that wielding it would be so lonely. As he passed through London for the last time in 1828, the metropolis that had once seemed to offer infinite variety now appeared empty of promise. 'My health was gone, my ambition was satisfied, I was no longer excited by the desire of distinction', he recorded bleakly in *Consolations of Travel*; 'my cup of life was no longer sparkling, sweet and effervescent; – it had lost its sweetness without losing its power'.[7]

Davy's self-destructive arrogance was manifested, too, in his relationships with his protégés, where he lacked Beddoes' natural aptitude in nurturing sons of genius. When he appointed the dissenter and blacksmith's son Michael Faraday to be his assistant at the Royal Institution in 1812, he had seemed to be replicating the choice of raw genius over social standing that Beddoes had made in his own regard; but where Beddoes, like John Borlase before him, had stood back to allow Davy free passage to the next stage of his career, Davy and Faraday's relationship ended in bitter wrangling over priority and plagiarism, and Davy's petulant and undignified campaign to block Faraday from membership of the Royal Society. He had seemingly learned little by the time of his death: John Tobin, thrilled by the prospect of exotic travels with a genius of science, found himself treated like a servant by a surly and monosyllabic companion who demanded to be read Shakespeare for hours on end without ever expressing the smallest flicker of satisfaction.

But it was not only Davy's experiences of the poison chalice of power that fed the bitterness of his last days: it was also the growing rejection of the vision of chemistry that he had brought to the world. His language of fluxes, forces and powers could be declaimed magnificently in the Royal Institution's lecture hall, but was difficult to reduce to testable theory, and a new generation of chemists were beginning to ridicule it as pompous verbiage that did not so much open up new worlds as attempt to close the door to the more precise and materialist view of the world that was emerging, stripped of the mystification of imponderable life forces. Davy, like Coleridge, held fast to the insistence that 'the necessary consequence of materialism is atheism':[8] he continued to insist that life could not simply be reduced to a property of matter, and to give scientific credence to the view that some vital spark or divine principle was required to make sense of the phenomena of life. But now German chemists were demonstrating that organic compounds such as urea could be synthesised in the laboratory without the addition of any mysterious life force, and John Dalton's theory of atomic weights, bonds and compounds was proving to have more explanatory power than Davy's high-flown vitalism. His new world was coming to look like a reactionary outpost of science: as Erasmus Darwin might have put it, a feather bed to catch a falling Christian.

The decline in Davy's reputation would prove irreversible. After their public disputes over priority, Faraday refrained from further electrochemical experiments until after Davy's death; but when he resumed them, the thrust of his work was to mathematise Davy's mystical forces, refining them into the electrical field theory that opened the path to the invention of the dynamo and the electric motor. It would be Faraday's name, and not Davy's, that would be attached to the laws, theories, discoveries and inventions through which electricity would transform the world.

*   *   *

In many ways, the spirit of the 1820s had become far more sympathetic to Beddoes than to Davy. The Royal Institution was a harbinger of a new generation of projects that aimed to harness science for the public good. As Beddoes had urged Sir Joseph Banks, the state began to take centralised measures to quantify society's ills: infant mortality statistics were collected and the census expanded, and the results fed royal commissions that presented governments with ever more exact pictures of the state of the sick and drooping poor. Francis Bacon became a touchstone for the new Utilitarian movement, for which science and education were the key to practical advancement, and the word 'empirical' began to lose its pejorative associations with quackery and to acquire its respectable modern sense of observations validated not by theory but by experiment.[9] The Society for the Diffusion of Useful Knowledge, founded in 1826, was making the latest scientific knowledge available to working men and women hungry for the education that their parents' generation had been denied. The activists who had first mobilised to protest the slave trade and the Gagging Bills extended their crusades, expanding the reach of their charities and foundations into the slums, patching civil society more effectively into government and setting their sights with realism and determination on the previous generation's dreams of political reform.

Many of the causes and projects that drove this new spirit were those for which Beddoes, in the 1790s, had been a voice in the wilderness. Once disease and mortality statistics began to be assembled, it was clear that a significant factor was inadequate ventilation in the houses of the poor, and Evangelical and other charitable campaigns began to transform the nation's slums into healthier places to live. The vogue for liberal education swept through the wider population on a wave of moral sentiment towards the young that argued, as Beddoes had, that brutalised children grew up to become brutalising adults. The medical profession was reformed: the public conversation that Beddoes had joined with Sir Joseph Banks grew into the Apothecaries Act of 1815, whereby physicians were licensed and controlled, and chemistry classes became compulsory for medical students. The new generation of physicians turned to *The Lancet*, established in 1823 by the surgeon and social reformer Thomas Wakley, for waspish exposés not merely of quacks and mountebanks, but of the corruption that still riddled the highest levels of the professions. While Davy continued to reject the geological theories of James Hutton on the grounds that their materialism left no room for divine intervention, Charles Lyell was readying himself to prove the reality of the vast abyss of time that Hutton and Beddoes had recognised forty years before. And Lyell, in so doing, would set the frame within which Erasmus Darwin's grandson (who was also Tom Wedgwood's nephew) would reveal the origins of life in ways that Beddoes and his circle had glimpsed, but had been forbidden to speak of.

In his Hamlet frame of mind, the dying Beddoes had been convinced that his experiments had produced neither branch nor blossom nor fruit; yet in his more sanguine moments he had recognised with equal certainty that the process of deciphering the world was unstoppable. Twenty years after his death, his protégés continued to contest his influence on the new century. Davy, in a set of unpublished biographical notes, judged his former patron harshly: Beddoes had had ingenuity and raw talent, but both were betrayed by a 'wild and active imagination'. He was 'little enlightened by experiment and, I may say, little attentive to it', and as a consequence had ridden his hobbyhorses into the ground. He had, in Davy's mature view, 'lived too little among superior men' who might have restrained him from dissipating his energies in self-indulgence.[10]

In the *Encyclopaedia Britannica* of 1824, however, he was assessed far more generously, as a passionate campaigner in the nascent field of preventive medicine who had 'the imagination of a poet and could paint in the most vivid colours the sufferings entailed by disease'. He had been 'very justly characterised as a pioneer in the road to discovery', even though he was 'more active in exciting the labours of others' than in claiming credit for his own achievements.[11] The author of the article was Peter Mark Roget, who had become a successful physician in London and a popular lecturer on physiology and anatomy, and who would finally, after his retirement in 1840, find the time to complete the *Thesaurus* he had been planning since his youth. He also developed a series of mechanical inventions, including calculating machines and kaleidoscopes, and in 1825 established the principle of persistence of vision in a paper entitled 'Optical Deception in the Appearance of the Spokes of a Wheel Seen through Vertical Apertures'. It is a striking coincidence, if nothing more, that the essential principles of photography and the moving image were both discovered by Beddoes' sons of genius.

* * *

When Davy's ill-health and ill-temper forced him to relinquish the presidency of the Royal Society in 1826, the man elected to replace him was the one who had first set him on the path of science: Davies Giddy. He was now known as Davies Gilbert, having adopted his wife's surname on the death of his father-in-law, who had set this change as a condition for his inheritance of the considerable family estates; Gilbert became his *nom de guerre* for public life, though in private he preferred the authentically Cornish name of his birth. Giddy emerged reluctantly into the spotlight after a lifetime of public service behind the scenes. Across twenty years as an MP he had refused ministerial office on several occasions, but had distinguished himself as a pragmatic and tireless committee man, supervising public works and longitude competitions, weights and measures standards, commodity and agricultural policy, and lending his mathematical skills to the construction of steam engines and suspension bridges. He had long been a stalwart of the Royal Society, and indeed Sir Joseph

Banks had hoped he would succeed him to the presidency, but Davy's appetite for glory had made him an irresistible candidate, and Giddy's diffidence had found a better match with the backroom role of treasurer. Even in his most sanguine moments, Beddoes might have had difficulty in believing that the Royal Society would be presided over by two of his protégés in succession.

Throughout his demanding career and the blank formality of his marriage, Giddy had steadfastly honoured his promises to Beddoes, supporting Anna, supervising the education of her children, managing the trusts and rents that were left to them and, when they ran low, topping them up with his own funds. He took responsibility for the boys and in particular regarded the elder, Thomas Lovell, as his own son, funding his passage through Oxford University and setting him on the medical career that would be eclipsed by his macabre and ironic poetry and, in 1849, by his enigmatic suicide, apparently by means of curare poison. Giddy's attachment was made more poignant by the tragedies of his own offspring: his first child, a daughter, was born without the dura mater at the base of her skull and lived for seventeen years in a vegetative state, and his second son died at birth.

Anna migrated to the warmth of Italy for her old age, and by the time of her death in Florence in 1824 she had lived long enough to appreciate that Giddy, faced with the choice between love and duty, had chosen for the best. Their friendship had outgrown her infatuation, and in taking on the guardianship of her children he had given her the security that had, perhaps, lain at the root of her desire.

* * *

The spring of 1829 found Davy in Rome, weak and bedridden, 'a ruin among ruins',[12] and struggling to finish the final dialogue of *Consolations in Travel*, while John Tobin and Davy's younger brother, John, summoned from Malta, procured torpedo eels from the market for dissection in the room next door. Once the manuscript was complete, Davy's sourness and irritability seemed to pass, and he drifted dreamily on doses of morphine, sometimes emerging with the conviction that he had already died, and quoting with approval the sentiments of Lord Byron's 'Euthanasia':

> When Time, or soon or late, shall bring
> The dreamless sleep that lulls the dead
> Oblivion! May thy languid wing
> Wave gently o'er my dying bed.

On 26 March he requested his favourite works be read to him: Shakespeare, *The Arabian Nights*, and Tobias Smollett's rollicking picaresque novel *Humphry Clinker*, with its opening scenes set in the spas and clinics of Hotwells. Yet something in Davy was still not ready to give up. His mind overflowed with ideas of a brilliance that his brother found impossible to reconcile with his feeble physical state, a case of 'lightning before death'.[13] His health began to improve, and he insisted on a journey to the mountains and lakes of Geneva before the heat in Rome built to its intolerable pitch. For a month they travelled, stopping in Siena and Florence, Turin and Susa; the day they arrived in Geneva, Davy ate a hearty fish supper, looked longingly across the lake, speculated on the fly-fishing and went to bed. In the hour before dawn, his life force and animal body separated.

There were many who attributed Davy's early demise, like Beddoes', to the experiments tried upon himself. His daring at the Pneumatic Institution had first crystallised his public image, and with his death the pneumatic researches took their place in a narrative of hubris and nemesis, and of the raging ambition that had brought him to the pinnacle of science and cast him down in his prime. Joseph Cottle's retrospective diagnosis that the 'destructive experiments, during his residence at Bristol', had 'beyond all question, shortened his days'[14] would cling to his legend just as it did to Beddoes'. He had begun his career recklessly inhaling gases that sent him to oblivion, croaking 'I do not think I shall die'; he had ended it waxing lyrical on the genius that was destined to make him immortal, but he had perhaps lived long enough to doubt whether it would really be so.

* * *

Two years after Davy's death the Cornish doctor John Ayrton Paris assembled his first biography, drawing heavily on his letters, notebooks and poems, and on the memories of his associates. But already, in 1831, the world of the Pneumatic Institution, and indeed the very idea of pneumatic therapy, had receded beyond the comprehension of the author and his readers. 'In physic, theory and experience are in open hostilities with each other', Paris offered by way of explanation. 'The gases are never now employed in the treatment of disease, except by a few crafty or ignorant empirics, whose business it is to enrich themselves by playing on the credulity of mankind . . . they are the cast-off clothes of philosophers, in which the rabble dress themselves.'[15] Paris' confident dismissal is a reminder of how little interest the medical profession was paying in the 1830s to the agents that, under the name of anaesthetics, would soon be hailed as the greatest contribution ever made to medicine. In the wider literary world, too, the pneumatic researches had fallen into obscurity. When in 1818 Thomas Love Peacock made the protagonist of his *Nightmare Abbey* 'the author of a treatise called Philosophical Gas, or a Project for the General Illumination of the Human Mind', he was reaching for an exemplar of cranky utopianism to parody the illuminatist schemes of his friend Percy Shelley.[16] Most who had once championed pneumatic medicine had died, and those who had survived preferred that their involvement be forgotten.

But Paris' reference to 'the rabble' is a reminder of the curious half-life enjoyed by nitrous oxide during its wilderness decades of the early nineteenth century. In the immediate wake of the Pneumatic Institution, a trickle of medical research had begun to flesh out Beddoes' and Davy's findings. A letter to *Nicholson's Journal* in 1802 noted that it reduced facial pain; by 1806 the experimenter from the Askeian Society, William Allen, had begun including it in his lectures at St Thomas' Hospital in London; in 1808 a student at the University of Pennsylvania, where the gas had become a philosophical party-piece, wrote a thesis tentatively proposing its use in pain relief. But hints such as these were not followed up. It was the antics and poetics of the Pneumatic Institution that had formed the defining image of the gas; Gillray's caricature of the philosophical party at the Royal Institution had overwritten Beddoes' and Davy's claims for its future medical applications. Nitrous oxide came gradually to find a makeshift home in theatrical and variety-hall performances, alongside mesmerism and other twilight wonders. In the process it acquired a new name which, to judge by the handbill for an evening at the Adelphi Theatre in London in 1824, was already familiar to the public by that time. Between an illusionist and a performance with musical glasses it was advertised that

The Nitrous Oxide, or
LAUGHING GAS
will continue to be administered to any of the audience who may choose to inhale it
The wonders of which were first experienced by
SIR HUMPHREY [sic] *DAVY*[17]

The name of the president of the Royal Society was useful for clothing the performance loosely in the raiments of scientific marvel, but the term 'laughing gas' indicated to the audience the spirit in which the experience should be taken. As it had in Dowry Square, the air-bag created a scenario in which the experimental subject was also the performer, invited to occupy the spotlight and well aware of the audience's expectations: spontaneous lunacy would be rewarded with applause, and modesty booed off the stage. A familiar handful of tricks for standing the gaff were assembled into a precursor of the stage hypnotism shows of today. Associates planted in the audience by the demonstrator would scoff at the alleged powers of the gas; challenged to take the stage, they would inhale it and leap around like maniacs; subsequent volunteers would follow suit. A visiting German, Dr Christian Schönbein, recorded his impressions of one of these shows: he was struck by how strongly their combination of philosophical earnestness and bathetic levity appealed to the British taste, and wondered if the gas was destined to diffuse more widely through society. 'Maybe', he speculated, 'it will become the custom for us to inhale laughing gas at the end of a dinner party, instead of drinking champagne'.[18]

This was far from what Beddoes had envisaged for nitrous oxide in 1799; and yet it would turn out to be the route by which the gas would eventually find the indispensable role in medicine that he had prophesied for it. Laughing gas performances found their greatest success in the carnivals and variety acts that ranged across the wide open spaces of America, where they could play a new town every night for years without ever retracing their path. Hucksters such as Samuel Colt, subsequently the inventor of the mass-produced revolver, toured a laughing gas act advertised with Robert Southey's phrase, 'the atmosphere of the highest of all possible heavens must be composed of this gas'.

It was within this seamy milieu that visiting physicians began to notice that those under the influence of the gas could stumble and injure themselves without feeling pain. In 1844 the dentist Horace Wells, after witnessing a chaotic laughing gas performance in Hartford, Connecticut, self-experimented with nitrous oxide for his own wisdom tooth extraction. He returned to consciousness with the prophetic declaration of 'a new era in tooth-pulling', but his subsequent public demonstration at Massachusetts General Hospital in Boston was an embarrassing failure, and the race to demonstrate pain-free surgery was won by Wells' former colleague William Morton with his mysterious volatile agent 'letheon', which turned out to be old-fashioned ether.[19] But by the 1860s there were perhaps a thousand mostly itinerant American dentists using nitrous oxide for painless tooth extractions, and it was reports of this thriving business that, in 1864, prompted the London dentist Samuel Lee Rymer to make successful use of nitrous oxide anaesthesia in Britain for the first time. The second trial took place on New Year's Eve of the same year, in Bristol General Hospital.

Few doctors offered more acute or harrowing descriptions of the pain of surgery before anaesthesia than Thomas Beddoes. In his *Rules of the Medical*

Institution for the Benefit of the Sick and Drooping Poor, he urges attention to the early signs of scrofula by warning that, if unattended, the lesions that it produces on the neck will eventually require an operation. He details unblinkingly the 'sand strewed to catch the falling blood – the sponges set to suck up that which oozes from the dissevered, palpitating flesh', and the surgeon 'hiding the knife he holds under his apron, whilst he begs the patient to draw up his nightcap over his eyes that he may not see the stroke'. It was a scene that Beddoes witnessed far too often for one who insisted that 'I would not put a human creature to a moment's pain but for its own good'.[20] Pneumatic medicine's picaresque journey from laboratory to lecture hall, variety palace to carnival to dentist's chair, and to the present where it is a fixture not merely in every dental surgery and operating theatre but in maternity wards, decompression chambers, casualty departments and ambulances, is one that its original projector could never have anticipated, but for which he would gladly have sacrificed priority and fame.

Notes

Prologue

1. Priestley 1792 p. 28.
2. *The Times*, 18/7/1791.
3. Priestley 1791a p. 4.
4. Priestley 1792 p. 30.
5. Priestley 1790 Vol. 1, p. x.
6. Burke 1790 p. 90.
7. Burke 1796 p. 26.
8. *Ibid.* p. 25.
9. Priestley 1790 Vol. 1, p. 1, p. xi.
10. *The Times* 15/8/1789.
11. O'Brien p. 209.
12. *The Times*, 30/6/1790.
13. *Ibid.* 18/7/1791.
14. Priestley 1791a p. 5.
15. Priestley 1791b.
16. Braithwaite p. 141.
17. Rose 1960 p. 83.
18. Schofield 2004 p. 277.
19. Letter to James Watt 19/1/1790, King-Hele p. 205.
20. Priestley 1790 Vol. 1, p. x.
21. *Ibid.*
22. Burke 1790 p. 90.

Chapter 1: Freedom's Garland

1. Southey 1807 p. 196.
2. *Ibid.* p. 197.
3. Rose 1960 p. 70.
4. *Ibid.* p. 76.
5. DG 41/41, 4/11/1791 Cornwall County Record Office.

6. Stock pp. 10–11.
7. Beddoes, Commonplace Book, Stock Appendix II.
8. DG 15, Davies Giddy Almanac 1791. In 1826 Davies Giddy (who by then had changed his surname to Gilbert) interleaved his diaries into a notebook and wrote comments and reminiscences alongside the original pages. 'Congenial muddling disposition' was his characterisation of Beddoes at this time, but with thirty-five years' hindsight.
9. DG Diary 16/2/1786, Todd p. 17.
10. Hutton 1788.
11. *Observations on the Affinity between Basaltes and Granite,* read at the Royal Society 27/1/1791.
12. DG 41, 4/11/1791.
13. DG 41/9, 2/11/1791.
14. Todd p. 34.
15. DG 42/21, 2/11/1803.
16. DG 41/6, 5/5/1791.
17. DG 41/9, 2/11/1791.
18. DG 41/6, 5/5/1791.
19. DG 41/15, 16/5/1791.
20. DG 41/1/1, n.d. (late 1791).
21. *Ibid.*
22. Stock p. 3.
23. *Ibid.*
24. Stansfield p. 23.
25. DG 41/10, 27/3/1791.
26. Beddoes 1791, Stock Appendix III.
27. All *ibid.*
28. DG 42, 2/3/1795. It should perhaps be noted that Beddoes made this claim in the context of a rival claim of priority from another physician, Thomas Trotter.
29. Beddoes 1799a pp. 7–11.
30. Beddoes 1804 pp. 13–17.
31. Beddoes 1799b p. 19.
32. Beddoes 1799a p. 287.
33. Beddoes 1799b p. 4.
34. Priestley 1774 Pt. II p. 102.
35. Beddoes 1793b p. 42.
36. Beddoes 1793c p. 29.
37. Vickers 2004 p. 32.
38. Yost pp. 101–2.
39. Quoted in entry for John Brown, *Oxford Dictionary of National Biography* (hereafter ODNB).
40. Beddoes 1793b p. 159n.
41. DG 41/41, 4/11/1791.
42. DG 41/9, 2/11/1791.
43. DG 41/29, 2/11/1791.
44. Beddoes to Black 23/2/1788, Levere 1981 p. 62.
45. Beddoes to Black 6/11/1787, Stansfield p. 37.
46. *A Memorial concerning the State of the Bodleian Library and the Conduct of the Principal Librarian Addressed to the Curators of That Library by the Chemical Reader* (probably 1787, see Stansfield p. 46).
47. Stansfield, p. 14.
48. DG 41/48, 21/11/1791.
49. DG 15, 27/10/1791.
50. DG 15, 19/2/1792.
51. DG 15, 12/5/1792.

Chapter 2: The Lunar Son

1. Beddoes to Joseph Black 15/4/1791 (ODNB).
2. Bodleian Dep. C. 134/2 (a), Reynolds to Beddoes 4/3/1789.
3. Dep. C. 135, Beddoes 5/6/1793.
4. Darwin 2007 p. xi.
5. King-Hele p. 47.
6. Stock Appendix 6.
7. Beddoes to Giddy Jan. 1792, Stock p. 32.
8. Journal of Katherine Plymley, Shropshire County Archives, 1066/5, pp. 17–20, 1066/6, p. 1.
9. DG 41/52, n.d.
10. *Ibid.*
11. DG 41/11, 6/3/1792.
12. Stock p. 43. The location is unspecified but is probably Kingsand, opposite Plymouth on the Cornish side of the Tamar estuary, where a vertically banded volcanic neck is exposed in a small disused quarry above the beach at Withnow.
13. DG 41/12, 11/4/1792.
14. Bodleian Dep. C 134/1 (a) 1.
15. Levere 1981 p. 62.
16. Beddoes to Black, *ibid.*
17. DG 41/11, 6/3/1792.
18. John Cooke 16/7/1792, Levere 1981 p. 64.
19. DG 15, 10/7/1792.
20. DG 41/35, 10/7/1792.
21. DG 41/14, 18/7/1792.
22. DG 15, 10/7/1792, n. 1824.
23. DG 41/14, 18/7/1792.
24. Beddoes 1792a p. 1.
25. *Ibid.* p. 2.
26. *Ibid.* p. 6.
27. Beddoes 1797b p. 59.
28. Beddoes 1792a p. 6.
29. *Ibid.* p. 9.
30. See Stansfield Ch. 5, n. 15.
31. Beddoes 1792b p. 4.
32. *Ibid.* p. 21.
33. *Ibid.* pp. 23–4.
34. *Ibid.* p. 32.
35. *Ibid.* p. 86.
36. *Ibid.* p. 79.
37. *Ibid.* p. 75.
38. Beddoes 1792c p. 109.
39. *Ibid.* p. 2.
40. *Ibid.* p. 89.
41. *Monthly Review* Vol. 20, 1796, p. 486.
42. Beddoes 1792d p. 1.
43. *Ibid.* p. 4.
44. *Ibid.* p. 5.
45. *Ibid.* p. 7.
46. *Ibid* p. 39.
47. DG 41/53, 12/1/1793.
48. DG 42/30, 14/3/1795.
49. DG 41/53, 1/7/1793.

50. Cunningham and Jardine p. 217.
51. Beddoes 1793b p. v.
52. *Ibid.* p. 1.
53. *Ibid.* p. 265.
54. *Ibid.* p. 114.
55. *Ibid.* p. 118.
56. *Ibid.* p. 44.
57. *Ibid.* p. 53.
58. *Ibid.* p. 147.
59. *Ibid.* p. 265.
60. *Ibid.* pp. 160–1.
61. Levere 1981 p. 65.
62. DG 41/19, 12/9/1792.
63. All *ibid.*
64. *The Times* 10/9/1792.
65. DG 41/56.
66. Beddoes 1792e.
67. All *ibid.*
68. Levere 1981 p. 66.
69. DG 41, 8/11/1792.
70. DG 41/22, spring 1793.
71. *Ibid.*
72. Beddoes 1793c p. 31.
73. All *ibid.* pp. 50–3.
74. Beddoes 1794b p. 3.
75. Darwin 2007 pp. 228–30.

Chapter 3: The Projector

1. DG 41, 8/11/1792.
2. DG 41/22, spring 1793.
3. All *ibid.*
4. Schofield 1963 p. 45.
5. Quoted in entry for Edgeworth, ODNB.
6. Stock p. 93.
7. *Ibid.* Appendix 8.
8. Wedgwood Mosley Collection 27.
9. Tom Wedgwood to his father, 7/7/1792, Litchfield p. 25.
10. *Ibid.*
11. Wedgwood p. 59.
12. Wedgwood, 'Etruria' and 'Liverpool' MS (hereafter W E/L) 28480–40.
13. Wedgwood p. 39.
14. *Ibid.* p. 39.
15. *Ibid.* p. 40.
16. Fissell p. 19.
17. *Ibid.* p. 159.
18. Beddoes 1793a p. 9.
19. *Ibid.* p. 20.
20. *Ibid.* p. 19, see DG 41/54, 8/10/1792.
21. DG 41/7, 3/7/1793.
22. To Robert Darwin, July 1793, Darwin 2007 pp. 93–5.
23. Beddoes 1793c p. 43.

24. DG 41/7, 3/7/1793.
25. Beddoes 1793c p. 10.
26. *Ibid.* p. 27.
27. *Ibid.* p. 44.
28. *Ibid.* p. 55.
29. DG 41/28, 15/6/1793.
30. DG 41/27, 26/5/1793.
31. All *ibid.*
32. DG 41/7, 3/7/1793.
33. DG 41/21, 26/5/1793.
34. Edgeworth to Mrs Ruxton, 21/7/1793, Edgeworth 1894 pp. 30–1.
35. DG 41/28, 15/6/1793.
36. Edgeworth 1820 Vol. 1 p. 153.
37. *Ibid.*
38. Bodleian Dep. C 135 (a), 5/6/1793.
39. DG 41/5, 8/11/1792.
40. Poole.
41. Rose 1793.
42. Harrison.
43. DG 41/4, 29/10/93.
44. Stock p. 100.
45. Foreman Ch. 17, n. 35.
46. Stock p. 100.
47. *Ibid.*
48. 4/3/1794, Stansfield and Stansfield p. 282.
49. Edgeworth 1820 Vol. 1 p. 155.
50. Bodleian Dep. C 134/7, 2/4/1794.
51. 14/6/1794, Stansfield and Stansfield p. 153.
52. 30/6/1794, *ibid.*
53. *Ibid.*
54. Wedgwood Mosley 35, early 1794.
55. Bodleian Dep. C 135 (a), 20/7/1794.
56. Beddoes 1794a.
57. All *ibid.*
58. Wedgwood Mosley 35, 12/8/1794.
59. Beddoes 1794b p. 10.
60. *Ibid.* p. 26.
61. *Ibid.* p. 40.
62. *Ibid.* p. 38.
63. Wedgwood Mosley 35, 7/11/1974.
64. DG 42/1, 31/11/1794.
65. Banks to Georgiana 30/11/1794, Edward Smith p. 189.
66. *Ibid.* p. 190.
67. *Ibid.* p. 191.
68. Levere 1977 p. 44.
69. James Watt jr to his father 5/9/1792, Robinson.
70. [Anon.] 1794.

Chapter 4: The Watchmen

1. DG 42/30, 14/3/1795.
2. DG 42/36, 12/2/1795.

3. Wedgwood Mosley 35, 6/3/1795.
4. DG 42/4, early 1795.
5. Wedgwood Mosley 35, 6/3/1795.
6. Wedgwood Mosley 35, 9/6/1795.
7. Gilbert 7/1/1795, Todd p. 26.
8. Wedgwood Mosley 35, 27/3/1795.
9. DG 42/6, 22/5/1795.
10. Beddoes 1794b 2nd edn.
11. DG 42/36, 12/2/1795.
12. *Ibid.*
13. Park.
14. To Horace Bedford 22/8/1794, Speck p. 46.
15. DG 41/42, 9/10/1793.
16. Cottle 1837 p. 154.
17. Charles Southey Vol. 2 p. 23.
18. Coleridge, 1956/2000 Vol. 1 p. 203, Vickers 1997 p. 52.
19. *Ibid.*, p. 96, 6/11/1794.
20. To his brother George, March 1798, Butler 1981 p. 80.
21. Stock Appendix VII.
22. *Ibid.* Stock gives the date of 'Domiciliary Verses' as 1795, although this has subsequently been questioned and a later date proposed – see Vickers 1997.
23. *Monthly Magazine,* 1 Oct. 1819 pp. 203–5.
24. Coleridge 1971, introduction.
25. Coleridge, 1956/2000 Vol. 1 p. 81, late Feb. 1795.
26. Cottle 1848.
27. DG 42/26, 16/6/1795.
28. *Ibid.*
29. DG 42/6, 22/5/1795.
30. *Sarah Farley's Journal,* Bristol [newspaper] 6/6/1795.
31. Goodwin p. 373.
32. Levere 1984 p. 194.
33. Veitch p. 326.
34. DG 42/23, 23/4/1796.
35. *The Star* 18/11/1795, reproduced as Appendix B1, Coleridge 1971.
36. All *ibid.*
37. Coleridge 1956/2000 Vol. 1, p. 93, 13/11/1795.
38. *The Star,* see note 35 above.
39. Coleridge 1971 Appendix B3, p. 373.
40. *Ibid.* p. 374.
41. *Ibid.* p. 375.
42. *Ibid.* p. 379.
43. *The Star,* see note 35 above.
44. All *ibid.*
45. Coleridge 1971 Appendix B4.
46. All *ibid.*
47. Coleridge 1970, introduction.
48. Stansfield and Stansfield p. 285.
49. Fullmer p. 85.
50. Beddoes 1795c.
51. Beddoes 1796a pp. 5–6.
52. *Ibid.* p. 6.
53. *Ibid.* p. 11.
54. *Ibid.* p. 12.

55. *Ibid.* p. 13.
56. *Ibid.* pp. 18–19.
57. *Ibid.* p. 22.
58. *Ibid.* p. 28.
59. Beddoes 1796b p. 51.
60. *Ibid.* pp. 197–8.
61. *Ibid.* pp. 198–9.
62. Coleridge 1817 Vol. 1, p. 177.
63. King-Hele p. 260.
64. *Ibid.*
65. Coleridge 1970 Vol. 2.
66. Coleridge 1956/2000 p. 10/2/1796, to Josiah Wade.
67. *Ibid.* p. 108, 21/3/1796, to Rev. John Edwards.
68. Cottle 1837 p. 156.
69. Beddoes 1796b pp. 9–10.
70. DG 42/20, 29/6/1796.
71. *Ibid.*
72. Stock Appendix 8.
73. 15/3/1796, Sandford p. 85.
74. Cottle 1837 p. 148.
75. Preserved in Joseph Cottle's album alongside pieces by Robert Lovell, Robert Southey and John Rose; Lamoine.
76. DG diary 31/5/1796, Todd 1967 p. 39.
77. *Ibid.*

Chapter 5: The Extraordinary Person from Penzance

1. Stock p. 134.
2. DG 42/33, 31/7/96.
3. *Ibid.*
4. DG 42/7, 23/8/1796.
5. Coleridge 1971 Appendix B3 n. 4.
6. DG 42/3, 5/9/1797.
7. Wedgwood Mosley 35, 3/8/1797.
8. Beddoes 1799a.
9. *Monthly Review* Vol. 20, 1796, p. 258.
10. DG 42/20, 29/6/1796.
11. Beddoes 1797a p. 10.
12. *Ibid.* pp. 15–16.
13. *Ibid.* p. 18.
14. *Ibid.* p. 70.
15. Litchfield p. 51.
16. Wordsworth Sept. 1806, *ibid.* p. 127.
17. Thompson p. 93.
18. Litchfield p. 54.
19. Beddoes 1797b p. 58.
20. Wedgwood Mosley 35, 3/8/1797.
21. DG 42/29, 17/10/1797.
22. Beddoes 1797b p. 12.
23. *Ibid.* p. 16.
24. *Ibid.* p. 20.
25. *Ibid.* p. 22.

26. Stock p. 144.
27. DG 42/15, 10/2/98.
28. *Ibid.*
29. Stock p. 150.
30. DG 43, 4/3/1803.
31. DG 42/26, 16/6/1795.
32. Paris p. 34.
33. W E/L 1548/2, 5/1/1799.
34. Paris p. 34.
35. 'Ode to St Michael's Mount', 1796, *ibid.* p. 23.
36. *Ibid.* p. 11.
37. W E/L, 1548/2, 5/1/1799.
38. Fullmer p. 88.
39. ODNB 1899, Vol. 60 p. 147.
40. Davy 1839 Vol. 1 p. 43.
41. Paris p. 18.
42. Hazlitt p. 217.
43. Wedgwood Mosley 35, 7/11/1794.
44. DG 42/28, 21/3/1798.
45. *Ibid.*
46. Davy 1839 Vol. 2 p. 82.
47. DG 42/2, 14/4/1798.
48. *Ibid.*
49. Beddoes to Giddy 4/7/1798, Paris p. 38.
50. 18/7/1798, *ibid.*
51. Fullmer p. 90.
52. Stock p. 151.
53. Clarke p. 166.
54. Edgeworth 1820 Vol. 1 pp. 185–6.
55. Stock p. 152.
56. *Ibid.*
57. Quoted in Stansfield p. 160.
58. *Ibid.*
59. Paris p. 43.
60. Davy 1839, Vol. 1, p. 43.
61. Cottle 1848 p. 263.
62. Fullmer p. 106.
63. Paris p. 52.
64. *Ibid.*
65. Knight 1992 p. 26.
66. Beddoes 1799a pp. 49–50.
67. 4/12/1798, Speck p. 76.
68. 27/1/1799, *ibid.*
69. Wordsworth 1970 Book X, l. 792ff.

Chapter 6: Wild Gas

1. Beddoes 1799b p. 6.
2. *Ibid.* p. 4.
3. Beddoes 1794b Appendix I (1796 edn).
4. Fullmer p. 96.

5. Paris p. 56.
6. W.D.A. Smith p. 22.
7. Davy's dates for his first experiments are self-contradictory. The date given in the published version of the *Researches* for the full experiment with Beddoes present is 17 April, but the same experiment was announced to Giddy in a letter of 10 April. In the *Researches* Davy makes mention of an earlier experiment, before the one that Beddoes attended; this sequence is confirmed by Beddoes' account. I take this earlier experiment to have been in the first week of April, and the letter to Giddy of 10 April, in which Davy refers to being 'absolutely intoxicated', to set a date of around 8 April for the experiment dated 17 April in the *Researches*.
8. Davy 1800 p. 458.
9. *Ibid.*
10. Beddoes 1799b p. 8.
11. Davy 1800 p. 459.
12. Stock p. 177.
13. W E/L 564–1.
14. Beddoes 1799b p. 14.
15. *Ibid.* pp. 19–20.
16. *Ibid.* p. 15.
17. Davy 1800 p. 464.
18. *Ibid.* pp. 508–9.
19. To Thomas Southey 12/7/1799. Stansfield p. 166.
20. Davy 1800 p. 346.
21. To Thomas Southey 12/7/1799.
22. Beddoes 1799b p. 12.
23. Davy 1800 pp. 499–500.
24. Beddoes 1799b p. 10.
25. Davy 1800 p. 522.
26. Cottle 1837 Vol. 2 p. 37.
27. Davy 1800 p. 297.
28. Beddoes 1799b p. 13.
29. Cottle 1837 p. 38.
30. Davy 1800 pp. 509–12.
31. *Ibid.* p. 496.
32. *Ibid.*
33. *Ibid.* p. 494.
34. W E/L, Notebook 3, 28515/150–40.
35. Beddoes 1799b p. 12.
36. Davy 1800 pp. 518–20.
37. *Ibid.*
38. Beddoes 1799b p. 10.
39. *Ibid.* pp. 16–17.
40. Cottle 1837 p. 37.
41. Davy 1800 p. 493.
42. *Ibid.* p. 493.
43. *Ibid.* p. 491.
44. Hoover p. 14.
45. Cartwright p. 238.
46. Beddoes 1799b p. 10.
47. Davy 1800 p. 501.
48. Beddoes 1799b p. 27.
49. W E/L, 1550–2.
50. Beddoes 1799b p. 30.

51. Davy 1800 p. 386.
52. *Ibid.* p. 465.
53. Cartwright p. 311.
54. Davy 1800 p. 476.
55. *Ibid.* p. 469.
56. Cottle 1837 p. 33.
57. Davy 1800 p. 469.
58. *Ibid.* p. 536.
59. 26/5/1799, Edgeworth 1894 pp. 165–6.
60. Davy 1800 p. 509.
61. Beddoes 1799b p. 34.
62. Cartwright p. 239.
63. Paris p. 71.
64. Davy 1800 p. 479.
65. Beddoes 1799b p. 4.
66. *Ibid.* p. 6.
67. *Ibid.* p. 7.
68. *Ibid.* p. 27.
69. *Ibid.* p. 33.
70. *Ibid.* p. 34.
71. *Ibid.* p. 45.
72. *Ibid.* pp. 36–7.
73. *Ibid.* p. 33.
74. Davy 1800 pp. 516–18.
75. Davy 1858 p. 74.
76. Letter to Southey, 17/12/1799, Coleridge 1956/2000 p. 303.
77. Schelling 1797, Cunningham and Jardine p. 31.
78. Coleridge 1817 Vol. 2 p. 272.
79. Coleridge 1956/2000 p. 527.
80. Davy 1839 Vol. 1, p. 119.
81. Burke 1757 p. 101.
82. Hazlitt p. 219.
83. 'Fears in Solitude', l. 12, Coleridge 1996.
84. *Morning Post* 12/12/1799.
85. Davy 1800 p. 479.
86. *Ibid.* p. 481.
87. *Ibid.* p. 483.
88. *Ibid.* p. 487.
89. *Ibid.* pp. 488–9.

Chapter 7: New Worlds

1. Poole p. 411.
2. DG 42/34, Dec. 1799.
3. W E/L 566–1, 12/11/1799.
4. DG 42/34, Dec. 1799.
5. DG 42/28, 21/3/1798.
6. Litchfield p. 36.
7. *Ibid.* p. 87.
8. *Ibid.* p. 89.
9. Cottle 1837 p. 41.
10. Cartwright p. 123.

11. Beddoes 1798 p. 2.
12. Stock p. 67n.
13. Cottle 1837 p. 41.
14. Paris p. 51.
15. Vol. 1, p. 90 (3rd edn).
16. *The Anti-Jacobin*, May 1800.
17. Davy 1800 p. 494.
18. *Ibid.* p. 558.
19. *Ibid.* p. 555.
20. *Ibid.* p. 559.
21. Paris p. 55.
22. Cartwright p. 235.
23. Fullmer p. 261.
24. *Ibid.* p. 279.
25. Davy 1800 p. xvi.
26. *Ibid.* p. 556.
27. Cottle 1837 p. 274.
28. 2/12/1800, Cartwright p. 312.
29. Elliotson p. 36.
30. Ball p. 187.
31. See Jay 2000 pp. 131–6.
32. Stent p. 84.
33. *Anti-Jacobin Review*, Aug. 1800, Article XI.
34. 13/5/1800, Litchfield p. 92.
35. 27/8/1800, *ibid.* p. 95.
36. Davy 1800 Appendix III.
37. Dibner Appendix A, p. 112.
38. Priestley 1767 Vol. 1 p. xiv.
39. Fullmer p. 285.
40. 20/10/1800, Paris p. 72.
41. W.D.A. Smith p. 30.
42. *Ibid.*
43. Paris p. 75.
44. Brown p. 122.
45. Davy 1839 Vol. 1 p. 132.
46. Fullmer p. 287.
47. *Ibid.* pp. 328–9.
48. 8/3/1801, Paris p. 87.
49. *Ibid.*
50. Fullmer p. 330.

Chapter 8: A Victim to Experiments

1. Speck p. 89.
2. DG 15, March/April 1801.
3. DG 89, 30/11/1800.
4. *Ibid.*
5. DG 89, 19/4/1801.
6. *Ibid.*
7. DG 89, 29/4/1801.
8. DG 89, 23/4/1801.
9. DG 89, 1/7/1801.

10. *Ibid.*
11. DG 89, 23/10/1803.
12. Knight 1992 p. 45.
13. Litchfield p. 188ff.
14. All *ibid.*
15. Treneer p. 79.
16. Cartwright 1952 p. 242.
17. Bence Jones p. 323.
18. Davy 1858 p. 64.
19. Beddoes 1804 p. 242.
20. Stock p. 300.
21. DG 42/8, 21/1/1802.
22. 12/3/1803, Cartwright p. 140.
23. Davy 1839 Vol. 1 pp. 322–3.
24. 27/11/1804, Levere 1982 p. 146.
25. Beddoes 1804 p. 2.
26. *Ibid.* p. 8.
27. *Ibid.* p. 9.
28. *Ibid.* p. 141.
29. *Ibid.* p. 249.
30. DG 89, 23/10/1803.
31. DG 89, 4/11/1803.
32. DG 89, 27/10/1803.
33. DG 89, Nov. 1803.
34. DG 89, 1/12/1803.
35. DG 15, 1/12/1803.
36. DG 42/21, 2/10/1803.
37. DG 42/24, 3/3/1803.
38. All *ibid.*
39. Paris p. 128.
40. Litchfield p. 179.
41. *Ibid.* p. 180.
42. DG 43/42, 9/5/1806.
43. DG 43/32, 7/5/1806.
44. DG 43/2, June 1804.
45. DG 89, 22/1/1808.
46. Beddoes 1808 p. 2.
47. *Ibid.* p. 5.
48. *Ibid.* p. 81.
49. *Ibid.* p. 82.
50. *Ibid.* p. 83.
51. *Ibid.* p. 87.
52. *Ibid.* p. 103.
53. *Ibid.* p. 124.
54. Stansfield p. 238.
55. DG 43/63, 9/11/1808.
56. DG 43/62, 9/11/1808.
57. Stansfield p. 239.
58. Cartwright p. 157.
59. DG 43/60, 24/12/1808.
60. Davy 1858 p. 106.
61. Coleridge 1961 3079.
62. Coleridge 2000 pp. 141–2.

Epilogue

1. Southey 1851 Vol. 1 pp. 338–9.
2. Sacks 1993.
3. Davy 1839 Vol. 1 p. 363.
4. *Ibid.* Vol. 9 p. 21.
5. *Ibid.* p. 34.
6. Davy 1858 p. 210.
7. Davy 1839 Vol. 9 p. 167.
8. *Ibid.* Vol. 1 p. 68.
9. 'An empirical law, then, is an observed uniformity, presumed to be resolved into simpler laws, but not yet resolved into them': John Stuart Mill, *Logic*, III, XVI (1846).
10. Papers Box 14 i, Royal Institution, Stansfield p. 249.
11. Supplement to the 4th, 5th and 6th editions, Vol. 2, Stansfield pp. 247–8.
12. Letter to Tom Poole 6/2/1829, Treneer p. 244.
13. Davy 1839 Vol. 1 p. 409.
14. Cottle 1848 p.40.
15. Paris p. 41.
16. Peacock p. 93.
17. W.D.A. Smith p. 34.
18. *Ibid.* p. 35.
19. Jay pp. 42–3, 132–7.
20. Beddoes 1804 p. 32.

A Note on Sources

After Beddoes' death, Anna commissioned John Stock, who had worked as his assistant during his final years, to write his biography, making available to him the chaos of papers, pamphlets, lectures and letters that had survived him. Both Davy and Coleridge had already declined the task, and Beddoes' reputation would be very different had either of them accepted it. Both foresaw that the outcome of Stock's labours would be a worthy but dull tome that would obscure Beddoes' life and work to posterity. Coleridge, when he heard the news of Stock's commission, wrote to Davy that 'I could not help assenting to Southey's remark, that the proper vignette for the work would be a funeral lamp beside an urn and Dr. Stock in the act of placing an extinguisher on it'.[*] John King, who worked closely with Stock, described him later as 'a literary undertaker'.[†]

Stock's *Memoirs of the Life of Thomas Beddoes M.D.*, published by John Murray in 1811, was everything his friends had feared, but it was the book for which Anna had hoped. Her concern had been to generate a respectable account of her husband's life that would favour his medical achievements over his controversial political views, a task for which Stock was well suited. A former republican himself, he had been caught up in an alleged revolutionary plot while studying medicine at Edinburgh in 1794, an event that had forced him to flee the country and had blighted his career for a decade. His version of Beddoes' life focuses

---

[*] Coleridge 2000 Vol. 3 p. 171 (30/1/1809).
[†] Letter in *Bristol Journal*, 30/9/1836.

doggedly on his medical career and publications, avoiding mention not only of politics but also of Coleridge, Southey, most of the Bristol circle, and the wider cultural hinterland in which his subject had moved. For much of its length it summarises his medical writings, ironing out their idiosyncrasies and replacing the vigour of their prose with bland paraphrase.

Stock's memoirs can, however, be supplemented by a wide range of primary sources that reveal Beddoes' role as a catalyst for the fusion of medicine and politics, science and poetry among the circle that assembled around him and the Pneumatic Institution. Over 150 letters to Davies Giddy, held at the Cornwall County Record Office in Truro, track his personal progress through the turbulent 1790s; more correspondence is held in the Bodleian Library in Oxford and the Wedgwood Archives at Keele University; he appears sporadically but often vividly in the published letters of Coleridge, Southey, Davy, James Watt, Joseph Black and Erasmus Darwin. Other first-hand witnesses to the Institution, such as Joseph Cottle and Maria Edgeworth, provide vignettes and anecdotes to supplement those dotted through Stock's account.

Despite the richness of these scattered sources, the effect of Stock's ponderous memorial was to obscure the diversity and energy of Beddoes' work and make his rediscovery more difficult. It was only in the 1980s that this task of recovery began in earnest; since then, a number of scholars have investigated various facets of Beddoes' work and explored its connections to the social, intellectual and political worlds in which he moved. I am particularly indebted to the researches of Roy Porter, Dorothy Stansfield, Trevor Levere, Neil Vickers, June Fullmer, A.C. Todd and F.F. Cartwright, on all of which I have relied at various points to untangle some of the knottier aspects of the story.

Due to an overrun on its refurbishment, the Royal Institution's holdings of Humphry Davy's notebooks and correspondence have been unavailable to researchers throughout the duration of this project. In their absence I have relied on the selections of letters, poems and notebook entries included in John Ayrton Paris' biography and John Davy's edited *Works*, and on further notebook extracts in more recent sources, particularly the work of June Fullmer.

Bibliography

Archive and Manuscript Sources

British Library

Moilliet, J.L. and Smith, Barbara M.D., *A Mighty Chemist: James Keir of the Lunar Society*. Privately printed 21 Oct. 1982

Bodleian Library, Oxford

Dep. C 134.1 (a) – Thomas Beddoes lectures, notes, drafts
    134.2 (a) – letters to Thomas Beddoes
    135 – letters to Thomas Beddoes and his sister Rosamund (to 1794)

Bristol Reference Library

Maby, Muriel G., *The Life and Letters of John King Esq., Surgeon*, MS B28753

Cornwall County Record Office, Truro

Davies Gilbert [Giddy] Archive
DG 15–16 – Almanacs 1791–5, 1796–1800
DG 40–43 – correspondence with Thomas Beddoes
DG 89–90 – letters, poetry, notes from Anna Beddoes

Keele University Special Collections

Wedgwood Mosley Collection:

Letters between John, Tom and Josiah Wedgwood 1790–1805
Thomas Beddoes letters to Tom Wedgwood
Letters to Thomas Beddoes from Erasmus Darwin, etc

Wedgwood 'Etruria' and 'Liverpool' Manuscripts:

Letters from Thomas Beddoes, Anna Beddoes, Davies Giddy
Tom Wedgwood – metaphysical essays and notebooks

Shropshire County Archives

Journal of Katherine Plymley, 1792–4, 1066/5–8, 27

Printed Primary Sources

[Anon.], *Strictures on a Pamphlet, Entitled 'An Impartial History of the Late Riots at Bristol'* (Bristol, John Rose, 1793)

[Anon.], *The Golden Age: A Poetical Epistle from Erasmus D—n MD to Thomas Beddoes MD* (Oxford F. & C. Rivington, 1794)

[Anon.], *The Riots at Birmingham, July 1791* (Birmingham, pamphlet repr. Arthur Bache Matthews, 1863)

Anti-Jacobin Review and Magazine, 'The Pneumatic Revellers: An Eclogue' (May 1800)

— 'Article XI: Notice of "Some Observations Made at the Medical Pneumatic Institution" by Thomas Beddoes M.D.' (Aug. 1800)

Bacon, Francis, *The Major Works* (Oxford University Press, World Classics revised edn, 2002)

Beddoes, Thomas, *Considerations on Infirmaries, and on the Advantage of Such an Establishment for the County of Cornwall* (Tredrea, 1 Sept. 1791)

— *Extract of a Letter on Early Instruction, Particularly That of the Poor* (25 Jan. 1792)

— *Alexander's Expedition down the Hydaspes & the Indus to the Indian Ocean* (London, J. Murray and James Phillips, 1792)

— *The History of Isaac Jenkins, and of the Sickness of Sarah his Wife and their Three Children* (Madeley, J. Edmunds, 1792)

— *Reasons for Believing the Friends of Liberty in France Not to Be the Authors or Abettors of the Crimes Committed in that Country* (Shifnal, 9 Oct. 1792)

— *A Guide for Self-Preservation and Parental Affection; or Plain Directions for Enabling People to Keep Themselves and their Children Free from Common Disorders* (Bristol, J. Murray; London, J. Johnson, 1793)

— *Observations on the Nature and Cure of Calculus, Sea Scurvy, Consumption, Catarrh and Fever* (London, J. Murray, 1793)

— *A Letter to Erasmus Darwin MD on a New Method of Treating Pulmonary Consumption and Some Other Diseases Hitherto Found Incurable* (5 July 1793)

— *On the Nature of Demonstrative Evidence* (London, printed for J. Johnson, 1793)

— *A Proposal towards the Improvement of Medicine* (Clifton, Bristol, 29 July 1794)

— *Considerations on the Medicinal Use of Factitious Airs, and on the Manner of Obtaining Them in Large Quantities* (Bristol, Bulgin & Rosser, 1794) (reprinted with Part 3, J. Johnson, 1795; with Parts 4 and 5, J. Johnson, 1796)

— *A Word in Defence of the Bill of Rights against the Gagging Bills* (Clifton, Nov. 1795)

— *Where Would Be the Harm of a Speedy Peace?* (Bristol, N. Biggs, 1795)

— *A Letter to the Rt. Hon. William Pitt, on the Means of Relieving the Present Scarcity and Preventing the Diseases That Arise from Meagre Food* (J. Johnson, 1796)

— *Essay on the Public Merits of Mr. Pitt* (J. Johnson, 1796)

— *Alternatives Compared, or, What Shall the Rich Do to Be Safe?* (London, J. Debrett, 1797)

— *A Lecture Introductory to a Course of Popular Instruction on the Constitution and Management of the Human Body* (Biggs & Cottle, 1797)

— *Suggestion Toward an Essential Improvement in the Medical and Surgical Departments of the Bristol Infirmary* (1798)

— *Essay on the Causes, Early Signs and Prevention of Pulmonary Consumption, for the Use of Parents and Preceptors* (London, Longman & Rees, 1799)
— *Notice of Some Observations Made at the Medical Pneumatic Institution* (Biggs & Cottle, 1799)
— *Hygëia: Essays Medical and Moral* (Thoemmes Continuum, 2004, introduction by Robert Mitchell [1802])
— *Rules of the Medical Institution for the Benefit of the Sick and Drooping Poor, with an Explanation of its Peculiar Design and Various Necessary Instructions* (Bristol, J. Mills, 1804)
— *A Letter to the Right Honourable Sir Joseph Banks, Bart. PRS, on the Causes and Removal of the Prevailing Discontents, Imperfections and Abuses in Medicine* (Richard Phillips, 1808)
Benger, E.O., Miss, *Memoirs of Mr. John Tobin* (Longman, Hurst, Rees, Orme and Brown, 1820)
Brown, John (ed. Thomas Beddoes), *The Elements of Medicine of John Brown M.D.*, 2 vols (J. Johnson, 1795)
Burke, Edmund, *A Philosophical Enquiry into the Origin of our Ideas of the Sublime and the Beautiful* (Penguin, 1998 [1757])
— *Reflections on the Revolution in France* (Penguin, 1986 [1790])
— *A Letter from the Rt. Hon. Edmund Burke to a Noble Lord* (London, T. Williams, 1796)
Coleridge, Samuel Taylor, *Lectures 1795 on Politics and Religion* (*Collected Works*, Vol. 1), ed. Lewis Patton and Peter Mann (Routledge and Kegan Paul, 1971)
— *The Watchman* (*Collected Works*, Vol. 2), ed. Lewis Patton (Routledge and Kegan Paul, 1970)
— *Notebooks* Vol. 1 ed. Kathleen Coburn (Bollingen, 1957)
— *Notebooks* Vol. 2 ed. Kathleen Coburn (Parthenon, 1961)
— *Collected Letters* Vol. 1, ed. Earl Leslie Griggs (Oxford University Press, 1956, 2000)
— *Collected Letters* Vol. 3 ed. Earl Leslie Griggs (Oxford University Press, 2000)
— *Selected Poems* (ed. Richard Holmes, Penguin, 1996)
— *Biographia Literaria* (Rest Fenner, 1817)
Cottle, Joseph, *Early Recollections; Chiefly Relating to Samuel Taylor Coleridge, during his Long Residence in Bristol*, 2 vols (Longman, Rees & Co., 1837)
— *Reminiscences of Samuel Taylor Coleridge and Robert Southey* (Houlston & Stoneman, 1848)
Darwin, Erasmus, *The Botanic Garden* (Joseph Johnson, 1791)
— *The Collected Letters of Erasmus Darwin*, ed. Desmond King-Hele (Cambridge University Press, 2007)
Davy, Humphry, *Researches Chemical and Philosophical, Chiefly concerning Nitrous Oxide, or Dephlogisticated Nitrous Air, and its Respiration* (London, J. Johnson; Bristol, Biggs & Cottle, 1800)
— *The Collected Works of Sir Humphry Davy*, ed. John Davy (Smith, Elder & Co., 1839)
— *Fragmentary Remains, Literary and Scientific, of Sir Humphry Davy*, ed. John Davy (John Churchill, 1858)
Edgeworth, Maria, *The Life and Letters of Maria Edgeworth*, Vol. 1 ed. Augustus J.C. Hare (Edward Arnold, 1894)
Edgeworth, Richard Lovell (concluded by Maria Edgeworth), *Memoirs of Richard Lovell Edgeworth Esq.* (London, R. Hunter, 1820)
Elliotson, John, *Numerous Cases of Surgical Operations without Pain in the Mesmeric State* (Lea and Blanchard, 1843)
Harrington, Robert, *The Death-Warrant of the French Theory of Chemistry* (London, W. Clarke, 1804)
Hazlitt, William, *Selected Writings* (Oxford World Classics, 1998)
Hutton, James, *Theory of the Earth*, in *Philosophical Transactions of the Royal Society of Edinburgh*, I Part II (1788)
Loutherbourg, P.J. de, *The Romantic and Picturesque Scenery of England and Wales* (Robert Bowyer, 1805)
Park, Thomas (ed.), *The Works of the British Poets, Vol. XLI* (London, J. Sharpe, 1808)
Peacock, Thomas Love, *Nightmare Abbey* (Penguin, 1969 [1818])
Priestley, Joseph, *The History and Present State of Electricity* (J. Johnson, 1767)

— *Experiments and Observations on Different Kinds of Air, &c*, 3 vols (Birmingham, Thomas Pearson, 1790, [6 vols 1774])

— *Letter to the Inhabitants of the Town of Birmingham* (1791)

— *The Duty of Forgiveness of Injuries: A Discourse Intended to be Delivered Soon after the Riots in Birmingham* (Birmingham, J. Thompson, 1791)

— *An Appeal to the Public on the Subject of the Riots in Birmingham* (Birmingham, Hillary & Barlow, 1792)

— *Memoirs of Dr. Joseph Priestley, Written by Himself*, edited and introduced by Jack Lindsay (Adams & Dart, 1970 [1806])

Rose, John, *An Address to the Citizens on the Present Melancholy State of Bristol* (Bristol, 1 Oct. 1793)

Schelling F.W.J., *Ideas for a Philosophy of Nature as Introduction to the Study of that Science*, translated by E.E. Harris and Peter Heath, introduction by Robert Stern (Cambridge University Press, 1988 [1797])

Southey, Charles Cuthbert (ed.), *The Life and Correspondence of Robert Southey* (Harper & Brothers, 1851)

Southey, Robert, *Letters from England* (Cresset Press, 1951 [1807])

Stock, J.E., *Memoirs of the Life of Thomas Beddoes M.D., with an Analytical Account of his Writings* (John Murray, 1811)

Wedgwood, Thomas, *The Value of a Maimed Life: Extracts from the Notes of Thomas Wedgwood*, selected by Margaret Olivia Tremayne, introduced by Mary Everest Boole (W. Daniel, 1912)

Wordsworth, William, *The Prelude* (Oxford University Press 1970)

Printed Secondary Sources

Articles and Essays

Badash, Lawrence, 'Joseph Priestley's Apparatus for Pneumatic Chemistry', in *Journal for the History of Medicine*, Vol. 19, April 1964, pp. 139–55

Barzun, Jacques, 'Thomas Beddoes, M.D.', in *A Jacques Barzun Reader* (Harper Collins, 2002)

Bergman, Norman A., 'Thomas Beddoes and Humphry Davy at the Pneumatic Institute', in *Anaesthesia History Association Newsletter*, Vol. 7, No. 1, Jan. 1989

Beyer, Werner W., 'Coleridge's Early Knowledge of German', in *Modern Philology*, Vol. 52, No. 3, Feb. 1955

Böhm, W., 'John Mayow and his Contemporaries', in *Ambix*, Vol. 11, 1963, p. 105

Cartwright, F.F., 'The Association of Thomas Beddoes, M.D. with James Watt, F.R.S.', in *Notes and Records of the Royal Society of London*, Vol. 22, 1967

— 'Pneumatic Medicine', in *Pharmaceutical Historian*, Vol. 1, No. 5, June 1970

Crosland, Maurice, 'The Image of Science as a Threat: Burke Versus Priestley and the "Philosophic Revolution" ', in *British Journal for the History of Science*, Vol. 20, 1987, pp. 277–307

Gibbs, F.W. and Smeaton, W.A., 'Thomas Beddoes at Oxford', in *Ambix*, Vol. 9, 1961, pp. 47–9

Grinnell, George C., 'Thomas Beddoes and the Physiology of Romantic Medicine', in *Studies in Romanticism*, Vol. 45, No. 2, 2006, pp. 223–50

Harrison, Mark, ' "To Raise and Dare Resentment": The Bristol Bridge Riot of 1793 Re-examined', in *The Historical Journal*, Vol. 26, No. 3, 1983, pp. 557–85

Hindle, Maurice, *Nature, Power and the Light of Suns: The Poetry of Humphry Davy* (Blackheath Poetry Society, 2002)

Hoover, Suzanne R., 'Coleridge, Humphry Davy, and Some Early Experiments with a Consciousness-Altering Drug', in *Bulletin of Research in the Humanities*, Vol. 81, No. 1, spring 1978

Jacobs, Maurice S., 'Thomas Beddoes and his Contribution to Tuberculosis', in *Bulletin of the History of Medicine*, Vol. 13, 1943

Leigh, Julian M., 'Early Treatment with Oxygen: The Pneumatic Institution and the Panaceal Literature of the Nineteenth Century', in *Anaesthesia*, Vol. 29, 1974, pp. 194–208

Levere, Trevor H., 'Dr. Thomas Beddoes and the Establishment of his Pneumatic Institution: A Tale of Three Presidents', in *Notes and Records of the Royal Society of London*, Vol. 32, No. 1, July 1977, pp. 41–9

— 'S.T. Coleridge: A Poet's View of Science', in *Annals of Science*, Vol. 35, 1978, pp. 33–44

— 'Dr. Thomas Beddoes at Oxford: Radical Politics in 1788–93 and the Fate of the Regius Chair in Chemistry', in *Ambix*, Vol. 28, Part 2, July 1981, pp. 61–9

— 'Dr. Thomas Beddoes: The Interaction of Pneumatic and Preventive Medicine with Chemistry', in *Interdisciplinary Science Reviews*, Vol. 7, No. 2, June 1982

— 'Dr. Thomas Beddoes (1750–1808): Science and Medicine in Politics and Society', in *British Journal for the History of Science*, Vol. 17, 1984, pp. 187–204

— 'Romanticism, Natural Philosophy and the Sciences: A Review and Bibliographic Essay', in *Perspectives on Science*, Vol. 4, No. 4, winter 1996, pp. 463–88

Miller, Albert H., 'The Pneumatic Institution of Thomas Beddoes at Clifton, 1798', in *Annals of Medical History*, New Series, Vol. 3, No. 3, May 1931

Poole, Stephen, *Popular Politics in Bristol, Somerset and Wiltshire 1791–1803*. Submitted as a PhD thesis to the Faculty of Arts, Department of Social and Economic History, Bristol University Sept. 1992

Robinson, Eric, 'An English Jacobin: James Watt, Junior, 1769–1848', in *Cambridge Historical Journal*, Vol. 11, No. 3, 1955, pp. 349–55

Rose, R.B., 'The Priestley Riots of 1791', in *Past and Present*, No. 18, Nov. 1960, pp. 68–88

Sacks, Oliver, 'The Poet of Chemistry': review of *Humphry Davy: Science and Power* by David Knight, in *New York Review of Books*, 4 Nov. 1993

— 'Scotoma: Forgetting and Neglect in Science', in *Hidden Histories of Science*, ed. Robert B. Silvers (Granta, 1995)

Smith, C.U.M., 'David Hartley's Newtonian Neuropsychology', in *Journal of the History of the Behavioural Sciences*, Vol. 23, April 1987

Smith, E. Brian, 'Humphry Davy, Thomas Beddoes and the Introduction of Nitrous Oxide Anaesthesia', in *Gases in Medicine: Anaesthesia* (Royal Society of Chemistry, 1998)

Stansfield, Dorothy A. and Stansfield, Ronald G., 'Dr. Thomas Beddoes and James Watt: Preparatory Work 1794–96 for the Bristol Pneumatic Institute', in *Medical History*, Vol. 30, 1986, pp. 276–302

Stent, Gunther, 'Prematurity and Uniqueness in Scientific Discovery', in *Scientific American*, Dec. 1972

Stern, Walter M., 'The Bread Crisis in Britain 1795–6', in *Economica*, New Series, Vol. 31, No. 122, May 1964, pp. 168–87

Sudduth, William F., 'The Voltaic Pile and Electro-Chemical Theory in 1800', in *Ambix*, Vol. 27, 1980, pp. 26–35

Temkin, Owsei, 'Basic Science, Medicine and the Romantic Era', in *Bulletin of the History of Medicine*, Vol. 37, No. 2, March–April 1963, pp. 97–129

Todd, A.C., 'Anna Maria, the Mother of Thomas Lovell Beddoes', in *Studia Neophilologica*, Vol. 29, No. 1, 1957

Vickers, Neil, 'Coleridge, Thomas Beddoes and Brunonian Medicine', in *European Romantic Review*, Vol. 8, 1997, pp. 47–94

Walker, Diana J. and Zachny, James P., 'Subjective Effects of Nitrous Oxide', in *Mind-Altering Drugs: The Science of Subjective Experience*, ed. Mitch Earleywine (Oxford University Press, 2004)

Yost, R.M., 'Sydenham's Philosophy of Science', in *Osiris*, Vol. 9, 1950, pp. 84–105

Books

Altman, Lawrence K., *Who Goes First? The Story of Self-Experimentation in Medicine* (University of California Press, 1986)

Ball, Philip, *The Devil's Doctor: Paracelsus and the World of Renaissance Magic and Science* (Arrow, 2007 [Heinemann, 2006])

Bell, Madison Smartt, *Lavoisier in the Year One: The Birth of a New Science in the Age of Revolution* (W.W. Norton, 2006)

Bence Jones, Dr H., *The Royal Institution: Its Founder and First Professors* (Longman, Green & Co., 1871)

Bird, Vivian, *The Priestley Riots, 1791, and the Lunar Society* (Birmingham and Midland Institute, 1991)

Braithwaite, Helen, *Romanticism, Publishing and Dissent: Joseph Johnson and the Cause of Liberty* (Palgrave Macmillan, 2003)

Bristow, Colin M., *Cornwall's Geology and Scenery: An Introduction* (Cornish Hillside Publications, 1996)

Brown, G.I., *Scientist, Soldier, Statesman, Spy: Count Rumford* (Sutton Publishing, 1999)

Butler, Marilyn, *Maria Edgeworth: A Literary Biography* (Oxford University Press, 1972)

— *Romantics, Rebels and Reactionaries: English Literature and its Background, 1760–1830* (Oxford University Press, 1981)

Cartwright, F.F., *The English Pioneers of Anaesthesia* (Bristol, John Wright & Son, 1952)

Clarke, Desmond, *The Ingenious Mr. Edgeworth* (Oldbourne, 1965)

Clow, Archibald and Nan, L., *The Chemical Revolution: A Contribution to Social Technology* (Batchworth Press, 1952)

Conrad, Lawrence (ed.), *The Western Medical Tradition 800 BC–1800 AD* (Cambridge University Press, 1995)

Crowther, J.G., *Scientists of the Industrial Revolution: Joseph Black, James Watt, Joseph Priestley, William Cavendish* (Cresset Press, 1962)

Cunningham, Andrew and Jardine, Nicholas (eds), *Romanticism and the Sciences* (Cambridge University Press, 1990)

Dibner, Bern, *Alessandro Volta and the Electric Battery* (New York, Franklin Watts, 1964)

Emblen, D.N., *Peter Mark Roget* (Longman, 1970)

Fisher, Richard B., *Edward Jenner 1749–1823* (André Deutsch, 1991)

Fissell, Mary E., *Patients, Power and the Poor in Eighteenth-Century Bristol* (Cambridge University Press, 1991)

Foreman, Amanda, *Georgiana, Duchess of Devonshire* (Harper Collins, 1998)

Forgan, Sophie (ed.), *Science and the Sons of Genius: Studies on Humphry Davy* (Science Reviews, 1980)

Fullmer, June Z., *Young Humphry Davy: The Making of an Experimental Chemist* (American Philosophical Society, 2000)

Golinski, Jan, *Science as Public Culture: Chemistry and Enlightenment in Britain, 1760–1820* (Cambridge University Press, 1999 [1992])

Gooding, D., Pinch, T. and Schaffer, S. (eds), *The Uses of Experiment* (Cambridge University Press, 1989)

Goodwin, Albert, *The Friends of Liberty: The English Democratic Movement in the Age of Revolution* (Harvard University Press, 1979)

Grant, William J., *Medical Gases: Their Properties and Uses* (BOC Medical, 1995)

Hartley, Sir Harold, *Humphry Davy* (Thomas Nelson, 1966)

Holmes, Richard, *Coleridge: Early Visions* (Hodder & Stoughton, 1989)

— *Coleridge: Darker Reflections* (Harper Collins, 1998)

Jay, Mike, *Emperors of Dreams: Drugs in the Nineteenth Century* (Dedalus Press, 2000)

King-Hele, Desmond, *Doctor of Revolution: The Life and Genius of Erasmus Darwin* (Faber & Faber, 1977)

Knight, David, *Humphry Davy: Science and Power* (Blackwell Publishers, 1992)
— *Ideas in Chemistry: A History of the Science* (Athlone Press, 1995)
Lamoine, Georges, *Notes on the Bristol Literary Circles 1794–8* (Institut des Recherches Interdisciplinaires de l'Université de Toulouse – Le Mirail 1973)
Lefebure, Molly, *Samuel Taylor Coleridge: A Bondage of Opium* (Quartet Books, 1977)
Litchfield, R.B., *Tom Wedgwood: The First Photographer* (Duckworth & Co., 1903)
Lorraine de Montluzin, Emily, *The Anti-Jacobins 1798–1800: The Early Contributors to the Anti-Jacobin Review* (Macmillan, 1988)
Lucas, F.L., *Studies French and English* (Cassell, 1935)
Meteyard, Eliza, *A Group of Englishmen (1795 to 1815), Being Records of the Younger Wedgwoods and their Friends* (Longmans, Green & Co., 1871)
O'Brien, Conor Cruise, *Edmund Burke* (Vintage, 2002)
Orsini, Giordano, *Coleridge and German Idealism* (Southern Illinois University Press, 1969)
Papper, E.M., *Romance, Poetry and Surgical Sleep* (Greenwood Press, 1995)
Paris, John Ayrton, *The Life of Sir Humphry Davy* (Colburn & Bentley, 1831)
Poggi, Stefano and Bossi, Maurizio (eds), *Romanticism in Science* (Kluwer Academic Publishers, 1994)
Pool, P.A.S., *The History of the Town and Borough of Penzance* (Corporation of Penzance, 1974)
Porter, Roy, *The Making of Geology: Earth Science in Britain 1660–1815* (Cambridge University Press, 1977)
— *Doctor of Society: Thomas Beddoes and the Sick Trade in Late Enlightenment England* (Routledge, 1992)
— *The Greatest Benefit to Mankind: A Medical History of Humanity from Antiquity to the Present* (Fontana Press, 1999)
— *Enlightenment: Britain and the Creation of the Modern World* (Allen Lane, 2000)
— *Quacks: Fakers & Charlatans in English Medicine* (illustrated, Tempus Publishing, 2001)
— *Flesh in the Age of Reason* (Penguin, 2004)
Reid, Helen and Stops, Sue, *On the Waterfront: The Hotwells Story* (Redcliffe Press, 2002)
Roe, Nicholas, *Wordsworth and Coleridge: The Radical Years* (Clarendon Press, 1986)
Rolt, L.T.C., *James Watt* (B.T. Batsford, 1935)
Sandford, Mrs Henry, *Thomas Poole and his Friends*, 2 vols (Macmillan & Co., 1998)
Schofield, Robert E., *The Lunar Society of Birmingham* (Oxford, Clarendon Press, 1963)
— *The Enlightened Joseph Priestley* (Penn State University Press, 2004)
Sedgwick, Sally (ed.), *The Reception of Kant's Critical Philososphy* (Cambridge University Press, 2000)
Shaffer, E.S., *Kubla Khan and the New Jerusalem* (Cambridge University Press, 1980)
Shea, William R. (ed.), *Revolutions in Science: Their Meaning and Relevance* (Science History Publications/USA, 1988)
Simpson, A.D.C. (ed.), *Joseph Black 1728–1799: A Commemorative Symposium* (Edinburgh, The Royal Scottish Museum, 1982)
Sisman, Adam, *Wordsworth and Coleridge: The Friendship* (Harper Press, 2006)
Smith, Edward, *The Life of Sir Joseph Banks* (John Lane, 1911)
Smith, G. Monro, *A History of the Bristol Royal Infirmary* (J.W. Arrowsmith, 1917)
Smith, W.D.A., *Under the Influence: A History of Nitrous Oxide and Ogygen Anaesthesia* (The Wood Library–Museum of Anaesthesiology, 1982)
Snow, Stephanie J., *Operations without Pain: The Practice and Science of Anaesthesia in Victorian Britain* (Palgrave Macmillan, 2006)
— *Blessed Days of Anaesthesia* (Oxford University Press, 2008)
Speck, W.A., *Robert Southey: Entire Man of Letters* (Yale University Press, 2006)
Stansfield, Dorothy A., *Thomas Beddoes M.D. 1760–1808* (D. Reidel, 1984)
Todd, A.C., *Beyond the Blaze: A Biography of Davies Gilbert* (Truro, D. Bradford Barton, 1967)

Thompson, E.P., *The Romantics* (Merlin Press, 1997)
Treneer, Anne, *The Mercurial Chemist: A Life of Sir Humphry Davy* (Methuen, 1963)
Uglow, Jenny, *The Lunar Men* (Faber and Faber, 2002)
Veitch, G.S., *The Genesis of Parliamentary Reform* (Constable, 1913)
Vickers, Neil, *Coleridge and the Doctors* (Oxford, Clarendon Press, 2004)
Wells, Roger, *Insurrection: The British Experience 1795–1803* (Alan Sutton, 1983)

Index

Lafayette, Marquis de 51, 83
Lake District 196, 227, 230, 254
Lamballe, princesse de 108
Lambton, John George 151, 229
  as 'Radical Jack', earl of Durham 160, 253
Lambton, William 150–1, 159–60, 163
The Lancet 256
laudanum *see* opium
laughing gas 260–1
  see also nitrous oxide
Lavoisier, Antoine 5, 8, 30, 35–8, 67, 106,
    152, 171, 180, 221, 246
  discovery of oxygen 36–8, 157
  theory of caloric 38, 154, 157–8, 223
Leslie, John 139
Lessing, G.E. 118
Levere, Trevor 278
Leyden jar 219
Linnaeus, Carl 37, 66
Liston, Robert 214
Liverpool 78
Liverpool, Lord (Robert Banks Jenkinson) 250
Locke, John 55–6, 61, 118
London 74–5, 102, 105–6, 111, 124–5,
    131, 133, 190, 196, 218, 222–5, 242,
    254, 260
Longman, Thomas 190, 191
Looe 77
Louis XVI 22, 68, 76
Lovell, Robert 114–7, 128, 130, 138
Lunar Society 7–8, 43, 46, 55, 75, 80,
    97, 106
Lyell, Charles 256

Mackintosh, James 23
Marat, Jean-Paul 68, 71, 92
Marazion *see* Mount's Bay
Massachusetts General Hospital (Boston) 261
materialism 5, 109, 119–20, 193, 196,
    222, 255
  see also atheism
Mayow, John 30–1, 89, 215
Medical Institution for the Benefit of the Sick
  and Drooping Poor 238, 242, 245
Medical Pneumatic Institution *see* Pneumatic
  Institution
Mendel, Gregor 215
Michaelis, J.D. 118
Mill, John Stuart 56, 275n
miner's lung 24, 154, 236
Mitchill, Samuel Latham 143, 170–1, 173
Montague, Lady Mary Wortley 149
Monthly Review 101, 118, 146, 222

More, Hannah 63, 133
Morning Post 191–2, 196
morphine 258
  see also opium
Morton, William 261
Morveau, Guyton de 37
Moseley, Benjamin 233
Mount's Bay 21, 140, 152–6, 183, 189, 232
  see also Penzance; Giddy (home at Tredrea)
Muir, Thomas 93, 121

Naples 150
Napoleon Bonaparte 137, 160, 196
National Assembly (France) 22, 51, 52, 69,
    70, 76
National Convention (France) 71, 93
Naturphilosophie 193, 222, 255
  see also Schelling, Friedrich
naval mutinies 147, 200, 206
Nelson, Admiral Horatio 160
Nepean, Evan 71
  see also Home Office
Nether Stowey 138, 147, 161, 196
Newton, Sir Isaac 32, 56, 82, 155, 157, 179,
    192, 206, 219–20, 222, 251, 254
Nicholson, William 152, 220
Nicholson's Journal 171, 210, 220, 221,
    222, 260
Nightmare Abbey v, 260
nitric acid/oxide 143, 144, 170–1, 178, 185,
    208, 219, 221
nitrogen 66, 171, 177, 208, 222
nitrous oxide 143, 170–99, 205–17, 225,
    232–3, 236–7, 249, 250–1, 260–2
  see also anaesthesia; animal experiments;
    Askeian Society; Avon Gorge; Beddoes,
    Anna; Beddoes, Thomas (pneumatic
    researches; satirised; self-
    experimentation; works: *Notice of Some
    Observations Made at the Medical
    Pneumatic Institution*); Berthollet,
    Claude-Louis; Coleridge, Samuel
    Taylor; Colt, Samuel; Cottle, Joseph;
    Davy, Humphry (*Researches, Chemical
    and Philosophical*); Gillray, James;
    Hammick, Stephen; Mitchill, Samuel
    Latham; palsy; Pneumatic Institution;
    respiration; Roget, Peter Mark; Royal
    Institution; Schönbein, Christian;
    Southey, Robert; Southey, Tom; Tobin,
    James Webb; Wedgwood, Tom; Wells,
    Horace
North, Lord Frederick 51, 52

Acknowledgements

Many thanks to Sharon Messenger for her tireless help in tracking down scholarly journals and articles; to Liz le Grice at the Cornish Studies Library, Desmond King-Hele, Steve Poole and Antonio Melechi for their help with specific queries; to Terry McCarthy and Gabriel Scally in Dowry Square; to John Beddoes of the Thomas Lovell Beddoes Society (http://www.phantomwooer.org/) for his generosity with family history and materials; to Chris and Linda Woolf for logistical support; and to Richard Barnett and Hasok Chang for their valuable comments on the manuscript.

Thanks posthumously to John D. Mackeson, whose papers came into my possession during the early stages of writing this book. Mr Mackeson was (according to the correspondence among his papers) a retired lawyer from Bristol who spent several years in the early 1970s researching Thomas Beddoes, his family, and the local and political history of his Bristol milieu. His notes and transcripts represent a great deal of work, and I would be glad to return them to his family or estate.

Quotations from the Wedgwood Archives are by courtesy of the Trustees of the Wedgwood Museum, Barlaston, Stoke-on-Trent, Staffordshire, England.

Many thanks to my agents, Rowan Routh and Hannah Westland, and to Robert Baldock and Rachael Lonsdale and all at Yale, for their belief in this book and their skill and dedication in realising it to such a high standard.

Special thanks, as ever, to Louise Burton and Michael Neve, for their encouragement, patience and support throughout the process.